ROUTLEDGE LIBRARY EDITIONS:
THE GOLD STANDARD

Volume 8

THE OUTLOOK FOR GOLD

THE OUTLOOK FOR GOLD

CHARLES MORGAN-WEBB

Routledge
Taylor & Francis Group
LONDON AND NEW YORK

First published in 1938 by George Allen & Unwin Ltd

This edition first published in 2018
by Routledge
2 Park Square, Milton Park, Abingdon, Oxon OX14 4RN

and by Routledge
711 Third Avenue, New York, NY 10017

Routledge is an imprint of the Taylor & Francis Group, an informa business

British Library Cataloguing in Publication Data
A catalogue record for this book is available from the British Library

ISBN: 978-1-138-56184-7 (Set)
ISBN: 978-1-351-24702-3 (Set) (ebk)
ISBN: 978-1-138-57583-7 (Volume 8) (hbk)
ISBN: 978-1-351-27128-8 (Volume 8) (ebk)

THE OUTLOOK FOR GOLD

BY

SIR CHARLES
MORGAN-WEBB

LONDON
GEORGE ALLEN & UNWIN LTD
MUSEUM STREET

FIRST PUBLISHED IN 1938

PRINTED IN GREAT BRITAIN BY
UNWIN BROTHERS LTD., WOKING

CONTENTS

FOREWORD

THE first chapters of this book were written to an accompaniment of grave disturbances in the public mind as to the utility and value of gold. Financial circles in the United States were hotly discussing the necessity of reducing the price of gold from thirty-five to thirty dollars an ounce. The Government of South Africa was pressing insistently for the return of Britain to a gold standard. A general panic of gold de-hoarding was being reinforced by the involuntary de-hoarding of £70 millions of gold in one month (June 1937) from the French Exchange Fund. Gold was being sterilised in wholesale quantities by the Government of the United States.

The public were so full of uncertainty as to the present value and position of gold that an enquiry into "The Outlook for Gold" appeared to be a wantonly speculative enterprise. As the chapters proceeded towards their conclusion, the efforts of the French Government to impose a managed currency on the gold-minded French people helped to make confusion more confounded. The difficulty of keeping a single eye on the future while the present offered such manifold distractions and contradictions has been ever present.

The author was, however, greatly comforted by the comments of a friendly critic who read the work in typescript. His observations were to the following effect:

"I congratulate you on sitting so discreetly on the fence through the whole of the nine chapters. Like all prophets, you have backed the future both ways."

He cited the last paragraph of the book in support of his opinion, saying "If the price of gold goes up you will quote the first sentence of that paragraph triumphantly and say 'I told you so.' If the price of gold goes down, you will quote the second sentence and say with equal truth 'I told you so.' "

It is comforting to know that in the midst of so many distracting influences, the calm, judicial attitude implied in the phrase "sitting on the fence" has been preserved throughout the nine chapters of the book.

More serious criticism in advance has been directed against the assumption made at the outset that the price of gold is governed by the Law of Supply and Demand. Against this assumption it has been contended that the price of gold is fixed by governments. In particular, it has been pointed out that the present price of gold is fixed at thirty-five dollars an ounce by the United States Government, and that the dollar price fixes the price of gold in all the currencies of the world.

The reply to this criticism is that the fixing of the price of gold by governments does not destroy nor supersede the Law of Supply and Demand. That Law applied to gold affirms:

i. if the demand for gold exceeds the supply, the price rises;
ii. if the demand for gold is less than the supply, the price falls;
iii. if the demand for gold is equal to the supply, the price remains stable.

Governments fixing the price of gold utilise the third of the above manifestations of the Law of Supply and Demand. If they rate the services of gold at such a high level, they can stabilise the price of gold by exercising perpetual vigilance and by expending unlimited funds in equating the demand for gold to its supply. But the Law of Supply and Demand is operating intensively all the time to keep their noses to the grindstone. A momentary relaxation of control, a momentary hesitation in supporting the market, and the inexorable Law demonstrates its vitality. In the five years September 1931 to September 1936, the Law of Supply and Demand brought three strong nations to their knees—Britain in 1931, the United States in 1933, and France in 1936. Their respective fixations of the price of gold collapsed the moment their power, or their desire, to equate demand with supply ceased.

Moreover, it is not true that the dollar price of gold fixes its price in all the currencies of the world. It would only be true if the foreign exchanges are fixed as well as the dollar price of gold. The sterling price of gold is deliberately dissociated from the fixed dollar price of gold. It not only varies with the foreign exchanges, it goes to a premium above, or a discount below, exchange parity, in response to a hoarding demand or a de-hoarding panic.

An excellent description of the working of the Law of Supply and Demand, supplementing the dollar price of gold, is to be found in the articles on "The London Gold Market," in the *Economist* (Banking Supplement) for October 16, 1937, and "Gold and the Dollar," in the *Economist* for November 13, 1937.

The outstanding utility of gold is the confidence it gives in the future, whether to individuals, to banking systems or to governments. But governments have brought discredit on gold by requiring it to perform unnatural and impossible functions, and by attempting to destroy the individual demand which originally created the value of gold.

The panic de-hoarding of gold which accompanied the first chapters of this book has been succeeded by a strong hoarding demand for gold as it is going through the press. So violent an oscillation of public opinion with regard to gold in the short period of four months has brought home to the author his rash temerity in undertaking to discuss its future. But on the principle that present action is determined by our anticipations for the future, the book may help to interpret the present. Gold is the subject of violent prejudices, ranging from excessive adulation to passionate contempt. The attitude of the author from his position of detachment "on the fence" is equidistant from both extremes. His approach has been strictly utilitarian, regarding gold, not as a fetish, not as a superstition, but as a substance that satisfies the very human instincts, pride of possession, and assurance for the future, for the majority of mankind.

C. M.-W.

October 10, 1937

CHAPTER I

SUPPLY AND DEMAND

"Dr. Freud relates that there are peculiar reasons, deep in our subconsciousness, why gold in particular should satisfy strong instincts and serve as a symbol. The magical properties, with which Egyptian priestcraft anciently imbued the yellow metal, it has never altogether lost.

"Almost throughout the world gold has been withdrawn from circulation. It no longer passes from hand to hand, and the touch of the metal has been taken from men's greedy palms. Gold is out of sight—gone back again to the soil. But when gods are no longer seen in their yellow panoply walking the earth, we begin to rationalise them; and it is not long before there is nothing left."—Mr. J. M. KEYNES—*A Treatise on Money*.

IN the above passages it is suggested that the value of gold is derived partly from its supposed magical qualities, and that both the magic and the value of gold will shortly disappear. On the other hand, there is much reason to believe that the value of gold is determined in the same manner as the value of all other commodities is determined, by the operation of the Law of Supply and Demand. Human psychology enters very largely, it is true, into the valuation of gold. But it enters equally into the valuation of all other commodities. The valuation of food, of clothes, of houses, of equipment for recreation, of necessities as well as luxuries, is determined largely by the psychology of the people using or enjoying them, in creating the demand which forms one factor of their value.

The psychology and the preferences determining the value of gold are just as rational, and quite as economically justifiable, as the psychology which determines the value of houses, or of furniture, or of cricket bats, or of government bonds. There is no justification for declaring that the value of gold is due to magic or superstition, while holding that the value of gold-mining shares is due to the play of normal economic forces. It is widely improbable that governments, bankers and international financers would pay £7, or 35 dollars, an ounce for a substance the value of which is due to a declining superstition, and the utility of which is rapidly vanishing before the forces of reason and education in an enlightened age.

The price of gold in the future will be determined exactly as it has been determined in the past, and as it is being determined in the present, by the operation of the age-old economic Law of Supply and Demand. The demand of governments for the services of gold is an integral factor in determining its price. By controlling demands they can fix its price. They control, but do not supersede, the Law of Supply and Demand. If the demand for the services of gold is greater than its supply, then the price of gold will rise. If the demand for its services should fall below its supply, its price will fall.

Gold rose to its position of distinction as the most

important of the precious metals because, in the long-distant period of the dawn of civilisation, the demand for its services exceeded its supply. It continued its career throughout the ages as the leading precious metal because it has provided a unique series of services for mankind which no other substance could have furnished, and which produced a demand for gold in excess of its supply. It has raised its price in the past six years from eighty-five to one hundred and forty shillings an ounce, or, alternatively from twenty to thirty-five dollars an ounce, because its release from the servitude of the gold standard greatly enhanced its utility and increased the demand for its services. In its comparative position of freedom it exerts a far more beneficial influence on the money, the commerce and the industry of the world than when its artificial fixity of price acted as a barrier to obstruct industrial progress.

There is some reason to believe that the qualities which have caused gold to play such a prominent part in the history and progress of mankind in the past will continue to operate in the future. Its present value, as measured by its price, is considerably greater than that of such commodities as steel, or copper, or rubber, or wheat. Their price is discounted by their profusion. The price of gold is enhanced by its scarcity. The scarcity of gold, added to its other qualities, enables it to perform a unique series of valuable services that no other substance can perform. It provides a rare and special kind of utility, which appeals with peculiar intensity both to the majority of individuals who comprise mankind, and to the authorities responsible for providing the world with its monetary supplies.

The utility of gold is of a different quality from, and indeed of a higher and more subtle quality than, the utilities provided by steel, or rubber, or wheat. They satisfy our material needs; they minister to our immediate and pressing necessities. Their price is determined accordingly.

But gold is the only substance that can satisfy the insistent psychological demand for some medium that will guarantee security for the future. Other forms of wealth are subject to rust, decay, corruption, obsolescence, depreciation, and disintegration. Gold is indestructible. It is the two-fold combination of scarcity and durability that has conferred on gold its unique position as the supreme medium for the storage of value for future requirements, thereby creating confidence in the future. Its price is determined by its efficiency in its performance of these pre-eminent services.

In his latest volume, Mr. J. Maynard Keynes has found it necessary to restore to Economic Science "the ancient distinction between the use of money for the transaction of current business and its use as a store of wealth." Among the causes which have operated to make economics largely a sterile pursuit, having nothing in common with modern methods of industry and commerce, is the neglect of the effect of money in the formulation of the Theory of Value. The classical

Theory of Value operating in a moneyless world is now being discarded as having no application to the economic world of reality. Such realities as the commercial importance of ready money, and the magnitude of the storage functions performed by modern forms of representative money, cannot be ignored by a Science professing to explain the mechanics of the production and distribution of wealth.

The development of numerous forms of representative money supplemented, without superseding, the crude hoarding of gold as a method of storing value. The financing of production and distribution in the modern world requires vast accumulations of liquid funds by governments, administrations, institutions and business corporations, considerably in excess of the hoarding of individuals. But though these liquid funds are not accumulated in the form of gold, the psychology of the banking and the business communities demands that they shall be guaranteed by substantial gold reserves.

The gold basis of all forms of representative money is dictated by experience, not by superstition. Gold has established itself as a desirable and acceptable commodity throughout the world. Its value is largely independent of the policies or the fate of governments, or of the effects of war. A few nations—Britain, the United States and the Scandinavian nations exhaust the number—are making timid and tentative steps towards efficient managed currencies to some extent independent of gold. But while money, in its short-term and short-range capacity as an internal medium of exchange, may be managed independently of gold, it is difficult to foresee the time when money in its wider functions, as a store of value or as an international currency, can be independent of gold in its valuable capacity as a unique material substance of universal acceptability.

There is no reason to believe that the majority of nations will ever have the desire to manage their internal currencies independently of gold. A managed currency implies a special and somewhat rare national psychology towards representative forms of money. This exceptional popular attitude towards money must be supplemented by a centralised and well-disciplined banking system. It is seldom that these two essential requirements for a managed currency are found together.

The United States after an experience of ten months on a managed currency (April 1933 to February 1934) reverted to a form of gold standard. The experience of France since the Three Power Monetary Agreement of September 1926 has demonstrated the intense difficulties of establishing a managed currency in a country whose people have a strong metallic complex so far as their money is concerned.

A striking confirmation of the limitation of the scope of the operations of managed currencies is afforded as this volume is passing through the press. The City Notes of *The Times* for September 27, 1937, state:

There has lately been a resumption of gold hoarding by

private persons—another indication of the lack of confidence in managed currencies. In the past week a fair amount of gold has been bought by private persons as a protection against currency depreciation. The renewed flight from the franc is one cause of the resumption of hoarding, but by no means the only one, though the hoarding is chiefly on Continental account.

Even those countries approaching towards an internal medium of exchange independent of gold, make no pretence that an international currency, or a currency utilised largely as a store of value, can be operated independently of gold. Representative money—bank notes, deposits, short-term money—is expressed in terms of national currencies, having only a limited sphere of operation. In so far as they are used for the storage of value or for financing international transactions, they must be guaranteed, both in time and space, by some universally acceptable medium. The link between a national currency and such a universal medium need not necessarily be rigid, as effected by a gold standard. Modern experience suggests that a degree of elasticity in the connection between a national currency and the international monetary medium is necessary to safeguard internal prosperity from external disasters and cataclysms.

The obsolescence and collapse of the gold standard do not necessarily imply that gold is incapable of performing valuable monetary functions. The gold standard was a special utilisation of the valuable monetary qualities of gold to bind the nations of the world in a rigid monetary structure. It presumed a comparatively uniform industrial rate of progress for all nations. It made little or no allowance for divergent standards of living, divergent national capacities, divergent rates of progress, divergent national ideals. The progressive nations forced the pace unduly for those that were more backward, and were themselves effectively retarded when they advanced at a rate more rapid than that achieved by the less progressive nations.

Suited to the moderately progressive pace of the nineteenth century, the gold standard was hopelessly incapable of coping with the intensely divergent national developments of the post-War period. But this failure to utilise gold to achieve a general uniformity of industrial progress throughout the world does not mean that gold has no monetary value whatever. Indeed, the history of the past six years has proved that there has been a much greater demand for the services of gold since it was released from the servitude of the gold standard. Gold as a commodity available at a variable market price, has performed more valuable monetary functions than when its price was rigidly confined between narrow limits.

Gold as a commodity in universal demand, freely exchanged at a variable market price, can perform all the functions required as a basis of an international currency—the settlement of trade balances, the financing of international loans, international movements of capital—and at the same time operate as a cushion to protect internal industry from adverse international occurrences.

The future of gold can be determined only by a consideration of the reality and the permanence of the demands for its services. If the demand for gold, apart from its use in industry and the arts, is based on superstition, its price will fall as education and enlightenment demonstrate the unreality of its claims. If it can perform real services, its price will depend upon the strength of the demand for its services in relation to its supply.

The reality of the demand for gold can be ascertained by a series of questions of which the following are examples: Is the past and present demand for gold for personal hoarding likely to be maintained in the future? Will the existing laws, administrative restrictions and prejudices against the hoarding of gold continue, or be relaxed? Is it possible for investment to satisfy fully the demand for the storage of individual savings? If investment cannot supply that demand, should gold be encouraged to satisfy it? Apart from the personal storage of savings, is there a large and growing demand for the storage of value in liquid funds for commercial, industrial, institutional and government purposes? Is a gold reserve necessary as a guarantee for the future value of liquid funds of great magnitude? Is a gold reserve necessary to finance international loans, trade balances and movements of capital? Are the proportions of gold reserves to liquid liabilities, accepted as adequate in the past, likely to be adequate in the future?

Difficult as these questions concerning the future demand for gold are to answer, they are paralleled by equally difficult questions as to its future supply. The testimony given to the Macmillan Commission in 1930, and the Macmillan Report itself issued in 1931, supply grave warnings against forecasts of the future supply of gold. Mr. Joseph Kitchen, the greatest living expert on the supply of gold, testified that the supply of gold would remain stationary for a time, and then fall, perhaps at an increasing rate.

Mr. J. M. Keynes, having a very shrewd suspicion that Britain would withdraw from the gold standard, and that, in consequence, the price of gold would rise, put the following question to Mr. Kitchen:

"If the value of gold were doubled, do you think that would appreciably increase the output of gold?"

To this question Mr. Kitchen answered:

"I should not think very much."

The value of gold has not doubled, and yet the output of gold has increased by 55 per cent since this evidence was given. How far the conclusions of the Macmillan Report have been falsified by experience may be judged from the two following extracts:

So far as the absolute output of new gold is concerned, the views of expert witnesses, both before our Committee and the Gold Delegation of the Financial Committee of the League of Nations, leave little room for doubt. The long-run tendency is for gold production to fall, though the rapidity of the decline in yearly output is differently assessed by different experts.

Also:

It must be concluded, therefore, that, so far as can be foreseen, the gold output of the world is likely to diminish.

The financial history of the past six years has demonstrated the falsity of such conclusions and the futility of all such speculations as to the future supply of gold. All that can be stated with certainty is that an increased demand for the services of gold will tend to increase its price, and that an increase in the price of gold will stimulate an increased supply by making it profitable to extract gold from sources which could not be made payable at the lower price. With equal certainty it can be stated that a diminished demand for gold will tend to lower its price, and that a fall in its price will be a deterrent to supply by rendering the less remunerative sources unworkable at a profit.

The cost of obtaining gold is, in fact, so great that supply waits on demand. It is on demand, rather than on supply, that the future of gold depends. If the money to finance the commerce and the industry of the world requires the services of gold (not necessarily in the form of a gold standard) for their efficient operation, and if the psychology of the majority of mankind regards gold as the most efficient medium for the storage of wealth for future eventualities, then the demand for gold will continue, and may indeed exceed its supply.

The only rational method of approach to investigate the probable future of gold is to consider the monetary demands for its services, the demand for industry and the arts being regarded as relatively stable. The possible monetary demands for gold are:

1. the demand for a medium of universal acceptability in which individuals who have not the aptitude, nor

the opportunity, nor the inclination, to utilise the method of investment, may store their life's savings;

2. the demand for gold as the basis of the currency of those nations which have not the aptitude to maintain a purely managed currency;
3. the demand for a medium of universal acceptability to link national paper currencies with each other and so enable them to perform international transactions;
4. the demand for a substantial material basis of general acceptability, independent of the prestige or stability of governments, to guarantee the future value of money, not in its capacity as a medium of exchange, but in its capacity as a store of value;

and the most insistent demand of all:

5. the demand for the restoration of their gold supplies to those countries from which it has been drained by adverse monetary influences in the past fifteen years.

CHAPTER II

THE STORAGE FUNCTION OF MONEY

"When a man is deciding what proportion of his money-income to save he is choosing between present consumption and the ownership of wealth. In so far as he decides in favour of consumption he must necessarily purchase goods—for he cannot consume money.

"But in so far as he decides in favour of saving, there still remains a further decision for him to make. For he can own wealth by holding it, either in the form of money (or the liquid equivalent of money) or in other forms of loan or real capital. The second decision might be conveniently described as the choice between hoarding and investing."—Mr. J. M. KEYNES—*A Treatise on Money.*

Two perpetual dilemmas have confronted mankind since the invention of metallic money. Each individual possessing money has repeatedly to consider one or both of the two problems:

Shall I spend, or shall I save?
If I save, shall I hoard, or shall I invest?

The early patriarchs, living a nomadic life, were spared the task of deciding the second of these problems. Their flocks and herds were both a form of money and investment. The use of cattle as money in the early days of civilisation is graphically brought to the consciousness of modern minds by the words "pecuniary" and "rupee," which are derived from the Latin and Sanscrit words for cattle.

Flocks and herds, besides forming a rough-and-ready medium of exchange, provided a valuable form of investment. The milk, flesh, hair, skin and wool they gave, provided a supply of food and clothing. They also provided a means of transport, and the natural increase in their numbers rendered them a valuable earning investment. Cattle as money performed a fourth function to the three usually assigned to money. They provided

a form of investment,

as well as performing the three normal monetary functions of providing

a standard of value,
a medium of exchange,
a store of value.

But, when mankind gradually withdrew from the nomadic life, and commenced to live a more settled existence, a new form of money was necessary. Cattle, sheep and goats could still be used as money by agriculturists, but not by the artisan dwellers in the cities and towns. In course of time, the metals, gold, silver and copper, demonstrated their superiority for performing three of the four functions hitherto performed by cattle in their capacity as money. Their rarity, their beauty, their durability and their general acceptability made them ideal substances to provide the best possible standards of value and media of exchange that could be obtained. But as stores of value, when compared with flocks and herds, they had one great disability. They were sterile. Flocks and herds were not only a store of value, they were an earning asset, an investment. Gold and silver, unless their treasured possession were sacrificed by being lent out on usury, could earn nothing. Hence the second of Mr. Keynes' dilemmas arose. Should the life's savings be stored in a sterile form by the accumulation of the precious metals? Or should they be stored in a form which would provide not only a store of value, but also function as an investment?

There was much to be said for both courses of action. On the one hand, the precious metals, in course of time, were found to possess the very valuable qualities of

> intrinsic desirability,
> general acceptability,
> stability of value,
> durability,
> personal physical possession,
> use as ornaments,
> ease of concealment and protection,
> immediate exchange value in any emergency.

Against these advantages they possessed in themselves no earning power.

On the other hand, the acquisition of flocks and herds, though they possessed the valuable quality of earning power, had many disadvantages. They were expensive and troublesome to look after and maintain in city surroundings; they were not generally acceptable to city dwellers as a medium of exchange; they could not be subdivided into small coinage; they were liable to mysterious diseases; they might die off altogether in a drought or an epidemic; and the pleasure and happiness of their actual possession made no strong appeal. In fact, they had most of the characteristics of a modern investment.

In course of time the advantages and disadvantages of both methods of storing savings became well known to all. There was no general rule applicable to all cases. Each individual decided for himself which method to adopt, in accordance with his opportunities, his environment and his personal psychology. Generally, investments in towns followed the same principles as investment by the possession of flocks and herds in the country. They were usually limited to the provision of tools, equipment and materials required in the personal pursuit of some occupation. The investment of savings in outside occupations, or investment in the purchase of land or real property, required special knowledge and skill, which few possessed, and constant supervision, which few could give. Moreover, they required larger sums of money than any but the wealthy could provide.

The great mass of mankind were accordingly limited to the method of storing all their savings, not required in the pursuit of their business or craft, in the form of metallic money, gold and silver for preference. Their psychology made them intensely jealous of the value of the hoard, so patiently and so painfully accumulated. They were fearful of entrusting their life's savings to anyone else for safe custody. Not only did they regard themselves as the best guardians of their store of savings; they derived considerable pleasure from the use of their stores of metals as ornaments. The value of their use as ornaments was enhanced by the knowledge that, owing to their general acceptability, they could always be used as money in an emergency. They possessed, in fact, the valuable quality of liquidity, or ready money. In one sense they were investments, giving an income of beauty, pleasure, adornment and the happiness of accumulation, instead of the more material returns derived from more formal investments. But gradually, in course of time, the use of gold and silver as a store of value, or hoarding, became dissociated from investment, and a distinct line of cleavage between hoarding and investment as a store of savings appeared.

As gold and silver established their supremacy as the two metals most suitable to perform the threefold functions of money of the higher denominations, a Standard of Value, a Medium of Exchange and a Store of Value, their attractiveness and utility for the particular function of being a Store of Value were greatly increased. They became more desirable, more acceptable, more liquid, more useful than ever in an emergency. The more gold and silver were demanded as the media for conducting operations of buying and selling by commerce and industry, the more valuable they became as stores of savings. They were never produced in sufficient quantities to supply the double demand for providing both a medium of exchange and a store of value. The shortage of supply relative to the constantly increasing dual demand for their services caused them continually to increase in value. They lost, to some extent, the quality of stability of value, but gained a much more popular quality to the hoarder, that of a steady appreciation of value, in comparison with the value of other commodities.

With the gradual increase of wealth, a certain amount of specialisation of function between gold and silver began to operate. They were both used as media of exchange. They were both used as stores of value. But silver established itself pre-eminently as the standard of value and the medium of exchange; and gold established an equal pre-eminence as a store of value. Gold was much too scarce and too precious to be risked in the rough-and-tumble of day-to-day transactions in the market-place. The operation of Gresham's Law enabled silver, the less valuable money, to drive gold, the best kind of money, out of circulation into storage. Gold was not needed except for transactions of an exceptional nature. In the main it remained aloof from the market-place, and became the most cherished possession of those whose savings were sufficiently great to enable them to acquire it, to use it for ornament and to store their wealth by its instrumentality.

The discovery of America transformed, to some extent, the nature of the problem involved in Mr. Keynes' second dilemma for those who sought a medium in which to store their savings, as expressed in the query:

If I save, shall I hoard or shall I invest?

The immense quantities of gold and silver poured into Western Europe increased their supply relative to the demand for their services, and transformed these metals from an appreciating asset into a depreciating asset. They arrived in such profusion that their supply exceeded the demand, and they lost two-thirds of their value, measured in goods, in the century elapsing between 1550 and 1650. Hoarders of gold and silver found their stores of savings rapidly diminishing in value. On the other hand, investors found their earnings greatly increased by the rapid rise of prices, and by the numerous openings for enterprise in the New World and in India and China. The balance of advantage between hoarding and investing as a method of storing savings was greatly modified. It tended to move from hoarding towards investment.

Nevertheless, though investment was established as a strong competitor, the tradition of hoarding as the outstanding method of storing savings was maintained. The opportunities for investment, except for the privileged and wealthy, were so few, that the hoarding of gold and silver remained, even after the investment boom of the Elizabethan era, by far the most widely adopted and the most prevalent method utilised for the storage of savings.

The rise of commercial banking in the eighteenth century, and the development of the use of various forms of paper money—Bills of Exchange, Commercial Bills, Treasury Bills, Cheques and Bank Notes—gave another strong stimulus to investment as a method of storing savings. It brought opportunities for investment within the reach of many who hitherto had been debarred from participation in the past. In particular, it attracted from hoarding to investment a large portion of the mass of previously stored money which was used to finance the English Industrial Revolution of the latter part of the eighteenth century. Though partly financed by bank credit, much of the capital used to finance industrial enterprise in Britain during the Industrial Revolution was supplied by the manufacturers themselves and their friends and connections from previous savings or from savings out of revenue. The Macmillan Report, in recording the method by which capital was provided to finance the Industrial Revolution, stated:

> Industry in those days was, so far as each unit was concerned, on a comparatively small scale; its basis was in the main a family basis; its capital was provided privately; and it was built up and extended out of profits.

Yet, despite the stimulus to investment caused by the Industrial Revolution, the hoarding of gold and silver still remained by far the predominant form of storing savings. The depreciation in value of these metals ceased in the latter part of the seventeenth century, and they regained their stability of value, the primary quality needed in a store of savings. The opportunities for investment still remained very restricted, available for the fortunate and privileged few in the more wealthy countries. The mass of the people in all countries still had no other method for the storage of their savings, beyond those incorporated in their own personal businesses, than the hoarding of gold and silver. The growing use of paper money for commercial and industrial purposes added to the glamour and popularity of gold and silver as stores of value. The insistence by economists and bankers that paper money derived its value from its gold and silver backing, added greatly to the prestige of these metals, and enhanced their value in the estimation of those having to make the decision as to the form in which to store their savings.

It was not till the nineteenth century and the development of international banking and commerce on the basis of the gold standard, that a transference from hoarding to investment as a store for savings was effected on a large scale. The call for development in all the backward countries of the world, the evolution of a more efficient international currency by the operation of the gold standard, and the increasing facilities for the issue of international loans, created an imperative demand for investment.

Simultaneously, numerous methods of collecting the innumerable stores of small savings of private individuals were developed in the more advanced countries. Joint Stock Companies, Savings Banks, Commercial Banks, Insurance Companies, Co-operative Societies, Friendly Societies, Building Societies, Trade Unions, all stimulated the habit of saving, and made the method of investment for the storage of savings available for all. For the period of roughly a century, from the establishment of the gold standard by Britain (1816) till the outbreak of the Great War (1914), the peoples of the more advanced countries of the world responded enthusiastically to the stimulus to utilise investment, rather than gold or silver, as the ideal medium for the storage of savings.

The increased use of banking facilities and paper money undoubtedly operated to diminish the attraction of the hoarding of the precious metals. There was an aesthetic pleasure attached to the possession of stores of gold and silver. Moreover, they could be used for ornament. There was no aesthetic pleasure, no gratification of the lust for possession, no pride of ornament, to be obtained from the possession of bundles of notes. Besides, banking provided a much safer and more reliable form of storing savings than hoarding. A deposit in a bank was much safer than a hoard of coins buried in the back garden. But a bank deposit gave little of the pleasure and happiness to be derived from adding slowly and patiently to a hoard that could be seen, and felt, and handled. If the private hoard was kept as a deposit in a bank at low interest, it might just as well be invested at the current rate of interest in a good security.

Thus, from the adventurous days of Queen Elizabeth till the outbreak of the Great War in 1914, the practice of hoarding was modified by a progressive movement towards the utilisation of investment as the best and most profitable method of storing savings and for making provision for the future. But this movement was of limited extent. Even in the more advanced and progressive nations it was not universal throughout all sections of the community. But in the less advanced nations, including some of those usually classified as Debtor Nations, the substitution of investment for hoarding as the accepted method of storing savings made little or no headway. Their peoples would not apply their savings even to achieve the development of their own countries. They preferred to pay a substantial tribute of interest to the investors of other nations rather than deplete their hoards and apply their savings to develop their own national industries.

Even before the War, the bulk of the populations of the world had not discarded the primitive method of storing their savings by hoarding gold, or silver, or money, for the alternative method of investment. But since the War, the movement away from hoarding towards investment has been checked, and, on the whole, reversed. The unsettlement caused by the War; the difficulties of post-War readjustment and reconstruction; the collapse of the gold standard; the numerous financial debacles, especially those of 1929, 1931 and 1933 affecting the two leading industrial countries, Britain and the United States; the rise of Fascism and Communism; the precarious post-War international situation; the great reduction, almost the cessation, of international lending; have all tended to destroy the previous confidence reposed in investment as the most secure and profitable method of ensuring provision for the future.

The loss of the prestige of investment as a method of storing savings has effected a complete transformation in the opinions of economists as to the functions of money. Prior to the War, the function of money as a Store of Value had been ignored and eliminated from consideration in textbooks on Economics. After the War, economists found themselves diverging further and further from reality by regarding money merely as an unreal symbol of wealth, incapable of performing the function of storage, instead of regarding it as it is in the actual world of industry and commerce, as the most potent form of real wealth in existence. It needed the genius and the forceful personality of Mr. J. M. Keynes to restore the storage function of money to its rightful place in Economic Science. His first attempt failed, largely because the phrase he used to denote this storage function, "the Bearishness of the Public," was too barbarous to merit general acceptance. His second attempt, "Liquidity Preference," describes with such perfect clarity and precision the storage function of money which economists had been endeavouring to ignore and to conceal, that it has of itself secured widespread acceptance of his new and revolutionary Theory of Interest.

Liquidity Preference is a comprehensive term comprising the many forms in which the storage function of money operates in the modern world. Modern commerce and industry require monetary storage on a scale immeasurably greater than can be effected by the hoarding of gold and silver. The commercial storage of value is effected by many modern monetary devices, the principal of which are bank notes, deposits (both fixed deposits and current accounts) and various forms of short-term money.

But though the storage of value is now effected largely by means of representative money, and not in the form of gold, it is very far from being independent of gold. The growing appreciation of the storage functions of representative money is accompanied by a growing feeling that representative money requires a far stronger gold backing as a guarantee of its future value, than has been considered adequate in the past.

Monetary management has progressed so rapidly since the collapse of the gold standard in 1931, that in the more progressive countries a gold backing is not necessary for money with reference to its function as an internal medium of exchange. The function of money as a medium of exchange is a short-term function. The prestige and efficiency of the governments of a few modern industrial nations are sufficient to guarantee the stability of money during the short periods occupied by the actual transactions of sales and purchases.

But the function of money as a store of value, whether exercised in the form of credit or deferred payments arising out of monetary transactions, or in any other of its many forms, is a long-term function. It requires as the guarantee of its future value something more material and substantial than the prestige and efficiency of governments. However great the experience, and however genuine the good faith of a government may be, it is liable to unforeseen and unpredictable changes of policy and personnel, and to such calamities as wars, loss of confidence and loss of financial credit. The worlds of finance, industry and commerce are not content that the large proportion of their wealth stored in money shall be entirely dependent on the prestige, the efficiency, and the stability of governments and of banking systems.

They demand, as a condition for continuing their activities, that the representative forms of money in which their immense liquid assets are stored shall be guaranteed by an adequate store of some universally acceptable commodity. International currency requires a similar guarantee. All forms of representative money have a limited area of acceptance. They are finally acceptable only within the jurisdiction of the authorities which authorise their issue. If they are to retain their value for the future, and to maintain their value in all countries, they need some material guarantee.

The experience of fifty centuries and the universal suffrage of mankind has selected gold to be that guarantee. It was the utility of gold as a store of value that was largely instrumental in giving gold its unique position as a precious metal. Its utility as a store of value is greater now than ever. It has not only to supply the still continuing heavy demand for individual hoarding in many countries, it has in addition to supply the reserves for the immeasurably greater volume of commercial, administrative and institutional monetary stores of value which are analogous to, but not identical with, primitive hoarding. It will also be in greater demand than ever as a war reserve in the future.

In a world of rapidly growing wealth the demand for provision for the future is constantly growing. Gold will be required in increasing quantities both for direct provision for the future in the form of personal hoarding, and also indirectly to guarantee the future value of the immense liquid stores of representative money which have superseded the hoarding of gold in the more advanced countries of the world.

At the moment of writing (June 1937) lack of confidence in the future value of gold has caused a severe panic resulting in a heavy de-hoarding of gold. It is essential to supplement this Chapter, describing the growth of hoarding from its primitive form, comprising the crude accumulation of gold, up to its modern manifestation in the respectable guise of liquidity preference, by a consideration of the factors which have been the cause of the heavy de-hoarding of gold for the past fifteen years.

CHAPTER III

THE DE-HOARDING OF GOLD

"All that the propensity of the public towards hoarding can achieve is to determine the rate of interest at which the aggregate desire to hoard becomes equal to the available cash. The habit of overlooking the relation of the rate of interest to hoarding may be part of the explanation why interest has been usually regarded as the reward of not-spending; whereas in fact it is the reward of not-hoarding.—Mr. J. M. KEYNES—Chapter 13, *The General Theory of Employment, Interest and Money*.

THE term "Hoarding" in the passage quoted above is used in its widest significance. It includes not only the hoarding of gold, but the operation of the storage function of money in all its forms—bank notes, current accounts or demand deposits, fixed or time deposits, and short-term funds of various kinds. It has the complete significance of the term "liquidity preference."

Similarly, the "not-hoarding" of the above quotation is something very different from the de-hoarding of the recent gold panic of June 1937. The not-hoarding of Mr. Keynes' revolutionary Theory of Interest includes much more than the relinquishment of the hoarding of gold. It includes the sacrifice of the possession of all and every form of ready money, representative money as well as gold. It means the sacrifice of liquidity preference.

The panic de-hoarding of gold during June 1937, was not de-hoarding at all. It was merely the transfer from one form of hoarding into another, from hoarding in gold to hoarding in other monetary forms. There has been no indication that the funds obtained from the de-hoarding of gold were utilised for the purpose of investment. On the contrary, they were transformed into the form of what is known in America as "hot money," that is, panic forms of stored money. In his speech, when moving the resolution to increase the Exchange Equalisation Fund by £200 millions in the House of Commons on June 28, 1937, Sir John Simon, the Chancellor of the Exchequer, though not actually using the term "hot money," described its characteristics in the following passages:

> Since the War, and especially since the suspension of the gold standard, capital movements of this sort (international) were far more frequent and far more sudden, and they were not necessarily directly economic or financial.

Also:

> Great quantities of short-term funds, which their owners might call back again when they pleased, were accumulated in the main financial centres, and thus, under the influence of, it might be temporary distrust or speculative ingenuity, they were shifted precipitately from one centre to another, and they got that situation which is sometimes picturesquely described as the movement of refugee capital.

And:

> Much depended for example on our short-term obligations; that was to say, on the amount of money which had been

placed in London from abroad and was liable to be recalled.

The recent action of the British and American Governments has demonstrated that the de-hoarding of gold has not reduced the demand for the monetary services of gold. Before Britain departed from the gold standard in 1931 it was considered that £150 millions was a sufficient gold reserve for Britain to hold. But in the speech mentioned, Sir John Simon revealed that on March 30, 1937, British gold reserves had risen in six years from £150 millions to £700 millions. On the date he was speaking they had approached £800 millions, and the object of his speech was to make provision to increase them by a further £200 millions. In the course of his speech, he stated:

> The conclusion which the Treasury was disposed to come to was that when all considerations were weighed they did not consider the present gold holding excessive.

In short, modern international and industrial conditions have changed the volume and the nature of the funds stored in representative forms of money, and particularly the funds known as "hot money," to such an extent that it has been necessary to increase the British gold reserves from £150 millions to a possible £1,000 millions.

The monetary authorities of the United States have been forced to increase their gold reserves by an even greater amount. All non-American monetary authorities regarded the gold reserves of the United States in 1929, amounting to £1,000 millions, as being excessive.

But since then the reserve requirements of the commercial banks of the United States have been doubled. The gold reserves have been increased from £1,000 millions to over £2,400 millions, and even at this figure doubts have been expressed as to whether they form an adequate safeguard against the avalanche of "hot money" pouring into the United States.

The de-hoarding of gold, whether public or private, does not reduce the demand for the services of gold. It increases the amount of "hot money" held in unstable monetary forms. It thereby necessitates additions to the gold reserves of the stronger industrial nations, held as safeguards against unpredictable monetary movements. Forced de-hoarding of gold in one direction necessitates forced accumulations of gold in other directions.

The de-hoarding of gold is seldom voluntary. It is usually forced upon the weaker of two competitors for its services. After the War, de-hoarding of gold was forced on the debtor nations by the gold accumulations of the United States under the national dollar-stabilisation form of the gold standard established by the Federal Reserve Board under Governor Strong. Subsequently, in 1928, France adopted a gold standard somewhat similar to that of the United States. The enforced de-hoarding of gold produced by the monetary policies of these two countries was strongly emphasised in the Report of the Macmillan Commission in 1931. The following passages from that Report are among the many references to these policies and their reactions on the debtor nations:

> The position of the debtor countries is apt to deteriorate rather than to improve as a result of their having to export gold.
>
> On the contrary, France and the United States have required payment of a large part of their annual surplus either in actual gold or in short-term liquid claims.
>
> The gold inflow of that year (1930) was fed by distress shipments from Spain, Argentina, Brazil and Australia.
>
> Failing this, the creditor countries will soon have sucked in all the gold available in the hands of the debtor countries.
>
> It is unlikely, therefore, that the debtor countries can continue much longer to square their international position by parting with gold at the rate of £70 millions a year, as they have been forced to do in the last two years.
>
> We repeat that it is the simultaneous reluctance of creditor countries (France and the United States) either to lend or buy, which is the cause of the crisis.
>
> Creditor countries must, unless they are ready to upset the economic conditions, first of the debtor countries and then of themselves, be prepared to lend back their surplus instead of taking it in gold.

The gold policies of the United States and France did more than force the debtor nations to de-hoard their gold. In 1931, they forced Britain to de-hoard gold until the gold reserves of the Bank of England were reduced to their last ounce. Britain did not leave the gold standard. The gold standard left Britain.

With the collapse of the gold standard in 1931, the de-hoarding of gold assumed a new form. The de-hoarding of the gold reserves of the debtor countries still persisted owing to the operation of the remnants of the gold standard in the countries of the Gold Bloc and in the United States. But this movement was supplemented by another form of forced gold de-hoarding, the de-hoarding of the stores of gold held by private individuals.

It is difficult for the people of Britain and the United States to appreciate the mentality of the peoples who still store their life's savings in the form of gold. In most advanced industrial countries, savings are almost automatically converted into investments. Liquidity preference is exercised to the extent necessary for commercial requirements; and, when exercised, is effected by means of representative money rather than in the form of accumulations of gold.

But in most countries, savings are not automatically converted into investments. The peoples of the less advanced countries have an extremely high liquidity preference and an extremely low degree of confidence in investment. This is demonstrated by their high rates of interest. If the new Theory of Interest propounded by Mr. Keynes is correct, and interest is the reward of not-hoarding, then their liquidity preference is intense. The rate of interest in many agricultural districts in India frequently rises to 6 per cent per month. This high rate of 72 per cent per annum is the measure of the liquidity preference, or the propensity to hoard possessed by the Indian cultivator.

This liquidity preference is expressed in terms of gold

and silver, principally gold. It is of an intensity that cannot be imagined by those who have not come into contact with it personally. It is not a recent creation. It is in accordance with the experience and the tradition of centuries. It has not been modified or weakened by the modern movement towards industrialisation. Capital requirements in countries with a liquidity preference for gold are met by loans from foreign countries.

This deep-rooted traditional form of hoarding has in the past few years been obstructed, and transformed into the actual de-hoarding of gold, by the following economic factors:

1. the heavy decline in the price level of primary commodities;
2. the high price of gold;
3. the disappearance of gold coins from circulation;
4. stringent restrictions on the possession and export of gold;
5. doubts as to the continuance of the high price of gold.

The heavy decline in the price of primary commodities, resulting from the deflation of the international currency, due to the operation of conflicting gold standards from 1925 to 1931, is the principal cause in the decline of gold hoarding. The practice of hoarding gold is most firmly established among the peoples of the countries producing primary commodities. With the continuous fall of wholesale prices commencing in 1925, their annual savings first decreased, then vanished, and were ultimately transformed into losses. The average prices of agricultural products in India declined by 53 per cent between 1926 and 1933.

Obviously there could be no fresh hoarding if there was no saving. As year after year passed with still declining prices, the precious hoards of gold were mortgaged to the local money-lender for the purchase of necessities, till better times should arrive. In 1931–32 the money-lenders preceived that the price of gold was advancing rapidly at a time when the prices of agricultural products were still falling. By foreclosing their mortgages they added to their gains by the sale of gold at a profit ranging from 50 to 66 per cent above its cost of acquisition.

The collapse of the gold standard in 1931 freed gold from the restrictions that limited its price to 85 shillings per ounce. The price of gold rose till it reached a level approaching 150 shillings per ounce.

The combination of extremely low prices for agricultural produce with a greatly enhanced price for gold transformed the traditional hoarding of many countries into de-hoarding. India, and to a less extent, China, afford the best examples of this transformation. From being continuous importers of gold to satisfy the insistent demand for hoarding, they were suddenly transformed into exporters of gold, this striking change demonstrating the heavy losses of those who would normally have been exporting their produce rather than gold.

The de-hoarding of gold from India and China from

1931 onwards is given in millions of pounds in the following statement:

	INDIA.	CHINA.	TOTAL.
1931	32	8	40
1932	68	12	80
1933	44	12	56
1934	47	8	55
1935	33	5	38
1936	25	4	29
Six Years' Total	249	49	298

The decline in the Indian and Chinese exports of gold from £80 millions in 1932 to £29 millions in 1936 does not indicate that their golden hoards are being exhausted. India has been hoarding gold from time immemorial, and the de-hoarding since 1931 has not exhausted even the 44 million ounces imported in the decade 1920–30. The rising prices of Indian and Chinese products since 1933 has lessened the amount of foreclosure by the money-lenders, and enabled the cultivators of primary products to save their cherished hoards of gold from sale and dispersal.

India and China are simply outstanding examples. The same process occurred in every country where the producers of primary commodities adhered to the primitive practice of hoarding their savings in gold. The declared policy of the British Government in raising the wholesale prices of primary products is gradually checking this form of de-hoarding. As the production of primary products becomes remunerative the capacity of their producers to save will be restored. It would be contrary to their mentality and their tradition for them to invest their savings. They will naturally resume their hoarding of gold with all the more fervour after their experience of the forced de-hoarding of their past savings.

The de-hoarding of gold in France, since 1935, is due to very different causes from those operating in India and China. The de-hoarding of the gold reserve of the Bank of France, by over thirty milliards of francs, from March 1935 to September 1936, was due to the artificially high value of the franc as a result of remaining on a gold standard.

The private de-hoarding of gold in France is also due to causes somewhat different from those working in India and China. It is due to administrative restrictions on the possession and export of gold, and to the disappearance of gold coins from circulation. So persistent is the hoarding instinct in France that, though the opportunity of hoarding gold has vanished, hoarding in the unattractive form of the accumulation of bank notes is most extensively practised. At the time of the first devaluation of the franc (September 1936) it was estimated that forty milliards of francs, or approximately half the note issue of the country, were being hoarded.

An even more striking demonstration of the French

passion for hoarding has been provided by the large volume of sterling pound notes hoarded on French account during the past two years. Rather than invest their savings, large numbers of French people prefer to hoard them in a foreign inconvertible currency. The extremely partial nature of the transference from hoarding to investment, effected in the nineteenth century, cannot be more clearly demonstrated than by this persistence of hoarding on a large scale in France, one of the leading industrial nations of the world.

The de-hoarding of gold in France from both private and public accumulations does not mean that France has lost the desire, or the will, to hoard gold. Owing to a faulty monetary policy for the five years, September 1931 to September 1936, France has gradually lost, for the time being, the power to acquire gold. Her de-hoarding of gold, like that of India and China, is forced and involuntary. Like the de-hoarding of India and China, it will be repaired as soon as the capacity to acquire gold is restored. As soon as the present financial difficulties of France have been overcome, it is reasonable to anticipate that she will seek to restore the £375 millions de-hoarded between March 1935 and September 1936, and the £100 millions Equalisation Fund, de-hoarded between September 1936 and June 1937, and also to transform her £400 millions of hoarded bank notes into gold. There is a potential demand for £875 millions of gold from France alone, to repair her involuntary de-hoarding, as soon as she obtains the resources with which to purchase it.

India, China and France are but three of the many countries suffering from the forced de-hoarding of their private and public accumulations of gold. Many countries—Germany is a conspicuous example—have been forced to adopt wasteful and expensive barter systems of international trade owing to the de-hoarding of their gold reserves. As a result of the forced de-hoarding of gold, few European or South American countries have adequate gold reserves to finance their internal and international monetary requirements. In the quotation from the Macmillan report, earlier in this chapter, it was stated that the forced de-hoarding of gold by the debtor countries in 1929 and 1930 was proceeding at the rate of £70 millions per annum, and the opinion was expressed that they could not stand the strain much longer. But during the past four years the de-hoarding of gold has been proceeding at more than three times that rate, at the rate of £223 millions per annum. The actual figures of de-hoarding of gold for the four years, 1933–36 inclusive, from the seven countries which have suffered most severely is as follows:

France	£513 millions
India	£148 "
Italy	£64 "
Germany	£63 "
Holland	£43 "
Switzerland	£31 "
China	£29 "
TOTAL	£891 millions

Thus the de-hoarding of gold, commencing in 1922, had reached an average of £70 millions per annum in 1930, and had risen to an average of £223 millions per annum for the four years 1933–36. It continued during 1937, and in one month, June 1937, the de-hoarding of gold from France alone amounted to £70 millions.

The de-hoarding of gold, so continuously proceeding on such an immense scale, is responsible for the following three menaces to the natural and orderly operations of industry and commerce:

1. the creation of "hot money," large numbers of persons being compelled to store their funds in forms of representative money, in which they have no confidence, instead of in gold, in which they have every confidence;
2. the adoption of prohibitive tariffs, exchange controls and restrictions, quotas and blocked and barter currencies, partly as defences against the drain of gold, and partly as monetary devices when the monetary store of gold has vanished;
3. the wholesale default of national and commercial obligations, debtor nations not having the gold with which to satisfy them, and not being able to pay in goods, owing to unremunerative prices and blocked currencies.

There is now ample production of gold to effect the restoration of their gold holdings to those nations from which they have been drained. There can be no general restoration of world prosperity until the forces which have caused, and are still causing, the de-hoarding of gold are controlled and reversed. Gold is the material basis of all forms of liquidity preference. The abandonment of the gold standard has not destroyed the monetary qualities of gold. It has set them free to assist in the financing of a far greater volume of commerce and industry than has ever yet been attained.

Fortunately, gold is now being produced in quantities sufficient for all monetary requirements. But a sufficiency of gold is not enough. It must be equitably distributed to perform its maximum utility, that is, to provide for each nation an adequate supply of money in those forms in which its peoples desire to transact their sales and purchases and to exercise their liquidity preferences.

CHAPTER IV

LIQUIDITY PREFERENCE

"Apart from the savings accumulated by individuals, there is also the large amount of income, varying perhaps from one-third to two-thirds of the total accumulation in a modern industrial community, such as Great Britain and the United States, which is withheld by Central and Local Governments, by Institutions and by Business Corporations for motives largely analogous to, but not identical with, those actuating individuals."—Mr. J. M. KEYNES—Chapter 9, *General Theory of Employment, Interest and Money*.

THE savings of a community may be divided into three categories:

1. individual savings, which are invested;
2. individual savings, which are held in some monetary form;
3. corporate savings, that is, monetary resources withheld from expenditure and investment by Central and Local Governments, by Institutions and by Business Corporations.

According to Mr. Keynes, the third category ranges from one-third to two-thirds of the total of the accumulated funds held in categories 2 and 3. The higher proportion is approached in the more advanced industrial countries; the lower proportion in the more backward countries. Indeed, the second and third of the above categories are largely complementary. In more backward countries, where individuals prefer to keep their savings in monetary forms, corporate savings are small. In the advanced industrial countries, where individuals prefer to invest their savings, it is necessary to hold large accumulations of corporate savings in monetary forms. Mr. Colin Clark, in his recent volume on *National Income and Outlay*, estimates that the corporate savings of Britain during 1935 were:

Undistributed profits	£228 millions
State and municipal savings ..	£83 ,,
Total	£311 millions

The important part which the storage function of money plays, not only in primitive communities, but also in the most advanced industrial communities, was ignored by most economists until it was forced upon their attention by the post-War monetary dislocations. Classical economics regarded money as being something outside the pale of reality, and as being merely a symbol or measure of reality, not to be regarded as possessing the attributes of real or material goods. They drew a wide distinction between real wages, real incomes, real goods, real capital, on the one hand, and money in its various forms on the other.

Economic Science evolved a Theory of Value from which money was excluded. The functions of Production and Distribution, and even those of Exchange, were presented as if they operated in a moneyless world. The practical business man, the industrialist, the merchant, who endeavoured to penetrate the mysteries of Economics, found themselves in a world as unreal as that in which Alice found herself in Wonderland.

In this Wonderland of the imagination, money did not matter, except frictionally or symbolically. It was denied all the attributes of reality or of wealth. It was merely a counter by means of which the value of the real goods and commodities essential for the welfare of mankind could be measured and exchanged.

Measuring instruments, such as yardsticks, scales, weights, clocks, sextants, theodolites, thermometers, providing the utility of measurement, were unhesitatingly included in the category of real wealth. But money, a measuring instrument which, as a Standard of Value, measured the price of every commodity on the face of the earth, was paradoxically excluded from the category of real wealth.

Instruments of exchange, such as shops, automatic machines, markets, banks, stock exchanges, were regarded as real facts, real capital goods, real wealth. But money, the supreme instrument of exchange, without which these subsidiary exchange instruments would have ceased to function, was denied admission to the category of real capital goods or real wealth.

Instruments of storage, such as warehouses, grain elevators, stock yards, oil tanks, gasometers, storing valuable commodities for future use, were regarded without question as real capital goods. But money, the traditional, age-old medium for the storage of wealth itself, was refused inclusion in the designation of real goods and real commodities comprising economic wealth.

To these important functions and utilities of measurement, exchange and storage, which made money a real and potent form of wealth, the power of the law added to money an additional and almost incalculable utility, that of legal tender. Legally, money was the only form of wealth which could cancel a debt. In virtue of this utility, this general acceptability enforcible by law, this legal conferment of the valuable quality of liquidity, or ready money, it is a more potent and living form of wealth in a modern industrial community than any other commodity that exists. And yet pre-War economists refused to regard it as having any reality as compared with other goods and commodities, or as possessing any of the attributes of wealth.

Two modern economists of such divergent outlooks as Wesley Mitchell and J. M. Keynes have found it necessary to expose the imaginative unrealities of the classical economists in their distortion of the nature of the operation of money in the modern world. Being concerned with such vital and practical problems as the Trade Cycle and Unemployment, these two economists have been compelled to maintain close contact with reality. Speaking of the necessity of dealing with actual business conditions, Wesley Mitchell, in his work on *Business Cycles*, states:

This conclusion runs counter to one of the traditions of economic theory. Most economists have explicitly subordinated the pecuniary aspect of behaviour, on the ground that money is merely a symbol, the use of which makes no difference save one of convenience.

Mr. J. M. Keynes expresses practically the same criticism in the passage:

> The conviction that money makes no difference except frictionally, and that the theory of production and employment can be worked out as being based on "real" exchanges, with money introduced perfunctorily in a later chapter, is the modern version of the classical tradition.

Under the misapprehensions that money is not wealth, that "money is merely a symbol," that "its use makes no difference except one of convenience," that "money makes no difference except frictionally," the function of money as a Store of Value was boycotted by economists and eliminated from the doctrines of Economics for a considerable period prior to the War.

Another misapprehension undoubtedly operated to obscure the function of money as a Store of Value. The fact that the amount of new investment is limited by the amount of saving (a truth dependent on the definitions adopted) was misapplied to produce the false conclusion that, as investment absorbed all savings, there could not possibly be a surplus of savings to be stored in the form of money. It was overlooked that whereas an individual had the limited choice between hoarding and investing his savings, the same limitation did not apply to a community. The equality of investment with saving is not an identity. The saving of one individual in money may be balanced by the investment of another individual in other goods. The amount of monetary saving is not limited or curtailed by the amount of investment. Savings in money and investment can advance concurrently to an indefinite amount. Monetary savings are limited only by the amount of money available (legal tender and other forms of money) to perform the various monetary services required by the community.

Whatever the reasons, the function of money as a Store of Value vanished from Economics for a lengthy period. The complete and absolute boycott of the storage function of money by pre-War economists can best be illustrated by a reference to the well-known and popular textbook of Mr. Hartley Withers: *The Meaning of Money*. Mr. Withers is an economist of international reputation and unimpeachable orthodoxy. His book on *The Meaning of Money* was intended to be an introductory compendium explaining in simple language for the use of students and beginners the operation of money, as conceived by contemporary economists. It passed through several editions. The opening sentence is as follows:

> The meaning of Money is not an economic theory. The object of this volume is to explain the matter of plain, positive, practical fact which is very important, very dull and very little understood.

Yet a reader would arise from a perusal of its 300 pages in complete ignorance of the "plain, positive, practical fact" that one of the most important functions of money from time immemorial has been the provision of a store of value, and that in a modern progressive industrial community the function of providing a passive store of value absorbs a far greater volume of money than is actively utilised as currency or as a medium of exchange. The function which money performs as a store of value is boycotted by an impenetrable veil of silence. A few incidental references in the book suggest that the utilisation of money as a store of value is a deplorable abuse of money. Such passages are:

> The public is much less likely to be bitten with the mania for hoarding its money.

And

> An outburst of hoarding on the part of both the public and the banks.

The boycott of the Storage Function of Money by economists persisted long after the War. Two such outstanding and progressive economists as Mr. R. G. Hawtrey and Sir Henry Strakosch still persist in excluding this function from the traditional and existing uses and purposes of money. Mr. Hawtrey in the first chapter of *Currency and Credit* imposes a limit on the operation of money in the following sentence:

> The use or purpose of money is two-fold; it provides a medium of exchange and a measure of value.

Sir Henry Strakosch, in his *Road to Recovery*, adopts the same limitation:

> They (the peoples of the countries on the gold bloc) seem to be impervious to the simple truth . . . that money is merely a measure of value and a medium of exchange.

Of course, such a dominating economic factor as the storage function of money could not be entirely excluded from consideration. It is introduced under disguised forms such as "the Consumer's Balance." In this form its volume and importance are belittled. It is regarded as a subsidiary factor to what is deemed to be the main function of money, that of being a medium of exchange.

The use of money for storage purposes was regarded (quite mistakenly) as a wasteful withdrawal of resources from the investment market, as a disastrous competitor preventing investment from contributing towards the progress and development of the world. The storage function of money remained under an economic cloud because it was regarded as the prime cause of trade depression in the following manner:

1. the volume of saving caused a glut of unsold consumption goods;
2. the storage of savings in monetary form curtailed investment and depressed the market for capital and construction goods.

The storage function of money is not, however, exercised at the expense of investment. It depends on the amount and kinds of money provided by the monetary and banking authorities for the operation of the two functions, the storage of value and a medium of exchange. If the amount of money is not sufficient or of the right kind to provide for the demand for storage, the rate of interest rises and investment suffers in consequence. But it is the shortage of money, or of the right kinds of money suitable for storage, not the demand for storage, that hampers investment.

Money is provided for two purposes: storage and exchange. If the perfectly legitimate demand for storage, in other words, liquidity preference, results in a high rate of interest for such monetary supplies as are available, the proper remedy is, not to ignore that demand as if it did not exist, nor to abuse it by applying to it such terms as "mania" and "outburst" (terms used by Mr. Hartley Withers), but to adjust the supply of the various kinds of money to the demand.

In a progressive industrial country it is not possible to provide legal tender to the exact amount of the demand for the services of money. The demand for money generally greatly exceeds the supply of legal tender money. Consequently, many expedients are devised in the normal course of trade to supplement the short supply of legal tender money by alternatives which will perform the functions of money as efficiently as legal tender. The general convenience of the commercial community renders such devices generally acceptable as money, subject to well-understood safeguards and regulations. They are not legal tender. Their purchasing power is conferred, not by law, but by tradition, by convention, by custom, and by the adoption of certain safeguards to guarantee the stability of their value and their acceptability at their face value.

These supplementary forms of purchasing power fall under the designations of Deposits and Short-Term Money. The term "Deposit" originally meant money actually deposited in a bank by its clients. But in course of time its meaning has changed, and it now comprises the large volume of representative money created by the banking system through its issue of loans and advances.

The designations, Short-Term Money or Short-Term Funds, indicate that although they are not legal tender they are regarded as being money by those who habitually make use of the various forms in which they appear. They comprise bills and loans of various kinds, usually for a period of three months or less. But bills and loans up to six months, and even twelve months, are frequently used for monetary purposes, and though not strictly included in the term "Short-Term Money" their uses are analogous to those of bills and loans for shorter terms.

The principal supplementary financial devices performing monetary functions are:

Current Accounts or Demand Deposits (circulating in the form of cheques),
Call Money,
Day-to-day Money,
Various terms of Short-Term Money ranging from one day to three months,

Fixed or Time Deposits of less than three-months term,
Three-months Treasury Bills,
Three-months Commercial Bills,
Three-months Bills of Exchange,
Fixed or Time Deposits of more than three-months term,
Bills or loans up to six-months term,
Bills or loans up to twelve-months term.

The earlier items of this sequence have a high degree of liquidity and earn either no interest or low rates of interest. Generally, the longer the term, the higher the rate of interest. As liquidity declines down the items of the sequence, the rates of interest rise. They are used primarily as money. Those which take the forms of short-term loans or bills are held in order to provide a constant succession of maturing credits, supplying day by day a volume of ready money sufficient for all anticipated demands. Their function in earning interest is secondary. The interest they earn pays something for their keep as they advance towards maturity to their ultimate utility as ready money.

It is sometimes contended that time deposits and short-term bills are not money. This raises the question whether it is possible to devise any practical and precise definition of money. Mr. J. M. Keynes gives a virtual negative to this question in his latest volume in the following passage:

> We can treat as money any command over general purchasing power which the owner has not parted with for a period in excess of three months, and as debt what cannot be recovered for a longer period than this; or we can substitute for "three months" one month, or three days, or three

hours, or any other period; or we can exclude from money whatever is not legal tender on the spot.

The essential quality of money is purchasing power. When used as a medium of exchange, the purchasing power of money is exercised. When used as a store of value, the purchasing power of money is held over for some future eventuality, either some definite payment or series of payments, or some remote, indefinite or unforeseen contingency that may arise in the far-distant future.

The difficulty that arises in framing a generally acceptable definition of the term "money" is due to the fact that its essential quality, purchasing power, is possessed by a whole range of financial devices and expedients, which, without being legal tender, perform the functions of money. Purchasing power is not confined to legal tender. In a barter economy, purchasing power is possessed to some degree by all commodities. And though, in a monetary economy, certain substances and documents are specially endowed with a special purchasing power by legal enactment or by custom or convention, this endowment does not completely deprive other substances and documents of their purchasing power.

In the absence of any feasible exact definition of money, for practical purposes money may be regarded as any generally acceptable substance or medium held primarily for its command over purchasing power, whether to be exercised immediately or in the future. Liquidity preference is the preference for holding wealth in a form providing the utility of purchasing power, rather than in forms which provide other utilities. The various forms of money shade off by almost imperceptible degrees into other forms of wealth as their liquidity declines and their capacity to earn low rates of interest, or to provide other utilities, appreciates.

The comprehensive nature of the storage function of money is best described in the following passage from Chapter 13 of Mr. J. M. Keynes' *General Theory of Employment, Interest and Money*:

> For the amount of hoarding must be equal to the quantity of money, or—on some definitions—to the quantity of money minus what is required to satisfy the transactions motive.

The first alternative, that the total volume of money in existence is employed in the function of storage, relies on the fact that an exchange transaction is momentary. It simply transfers money from one person's store of money to another's. In this somewhat theoretical sense all money is in storage.

A more practical view is that money employed in effecting a transaction in any one day is in active circulation, and that money not so employed is in passive storage. This view may be illustrated by considering the proportions of the money kept in current accounts or demand deposits, which are employed respectively in active circulation and in storage.

The common acceptation of the velocity of circulation of current accounts in Britain is about sixty per annum. Allowing for non-working days, this means that the money reposing in current accounts circulates on an average once in five days. This is equivalent to an average daily circulation of one-fifth of the total volume of the current accounts, and the permanent storage of four-fifths of their total.

It is popularly supposed that fixed or time deposits perform storage functions, and that current accounts or demand deposits are occupied in circulation only. But the slow circulation of the money contained in current accounts means that they are mainly occupied in performing storage functions, and that their daily transactions as a medium of exchange are effected on an average by one-fifth only of their total volume.

Thus, all existing money, except that small proportion financing transactions on any one day, is performing the function of storage. The duration of storage ranges from a few days, in the case of funds accumulated for such objectives as petty cash, or the payment of weekly wages, or the settlement of weekly accounts, to an indefinite period for funds accumulated for some remote eventuality. The totals of such stored money on any one day of the year, whether in coin or notes, or fixed or demand deposits, or in various forms of short-term money, approaches closely to the total of all money in existence, and, according to one possible definition, actually comprises all existing money.

The primary quality of this immense accumulation of stored money is liquidity, its potential capacity to be used as purchasing power, its quality of being ready money. The liquidity of legal tender money is conferred by law. The liquidity of such forms of representative money as deposits and short-term funds is a complex in which custom, convention and confidence in government and in the banking system are constituent parts. But these are not sufficient. Customs and conventions lose their authority in times of panic or crisis. Governments change. Banking systems have been known to collapse. In a dozen countries in Europe since the War, the purchasing power of immense and elaborate national systems of representative money vanished into nothingness.

In France the value of the wealth stored in its national monetary system was reduced in the proportion of tenpence to less than twopence. In the United States the monetary wealth of the country, according to the evidence of Governor Eccles, Chairman of the Federal Reserve Board, declined from 56 billion dollars in 1929 to 36 billion dollars in 1933, bringing down the annual income of the American people from 83 to 47 billion dollars.

To maintain the liquidity of representative money, something more than custom, and convention, and confidence in the integrity and efficiency of governments and banking systems is needed. There is one substance, which has maintained a high degree of liquidity and of universally extended purchasing power throughout the ages for fifty centuries, and still continues to maintain its liquidity and purchasing power in every country of the world at a uniquely high level.

There is only one substance in which the liquidity preferences of the peoples of all the countries of the world can be expressed. If customs and conventions collapse, if governments change, if banking systems suspend payment, there is one substance which does not lose its liquidity or its purchasing power, and which is the universal repository for the liquidity preferences of all mankind.

The liquidity preference embodied in gold being universal, it forms an ideal guarantee for the liquidity of all forms of representative money. This has been recognised by the universal provision of gold reserves as the basis for all such forms of money. It is demonstrated by the low efficiency of money in all countries where it is not effectively guaranteed by gold reserves.

The growing capacity for managing currencies affords no reason for abandoning the valuable services performed by gold in guaranteeing the liquidity of representative money. The capacity to manage currencies efficiently is very limited. A few nations only have shown this capacity. And even these few nations have not dared to depart from gold as the guarantee of the liquidity of their stored and international currencies. All that they have succeeded in doing is to make the short-term function of money as an internal medium of exchange independent of gold. But the purchasing power of money stored for future use, or of money required to finance international transactions, still requires to be guaranteed by substantial gold reserves.

Although gold has proved unequal to the task of guaranteeing the stability of the purchasing power of money, it still guarantees liquidity.

The increase of "hot money" or panic forms of representative money, calls for a much higher proportion of gold reserves than has hitherto been considered necessary. The main reason advanced by Sir John Simon in his proposal to increase the Exchange Fund by £200 millions was the monetary dislocation caused by this class of money. The £200 millions is required to augment the gold holdings of the Exchange Fund in order to guarantee commerce and industry from the unsettlement it causes.

As long ago as October 1933, a special article in *The Times* drew attention to the growing danger of panic money in the passages:

> The course of foreign exchanges is now governed mainly by the movements of short-term capital, capable of being transferred at short notice from one country to another. These ready-money balances were estimated two years ago to reach the sum of about ten billion dollars. The total mass of capital transferable at short notice is thus raised to a figure in comparison with which the normal trade balances of the world are insignificant, and which is far in excess of the total amount of monetary gold in existence. The existence of this enormous mass of ready money—which has no fixed policy of domiciling itself permanently in any one country, but runs from one to the other as prudence or panic dictates—makes it idle to suppose that exchanges can be stabilised by a general return to the gold standard.

Since the above was written, in 1933, the amount of panic money has been greatly increased by the de-hoarding of gold which has proceeded at an accelerated pace in the past four years. The countries which have lost their gold are precisely the countries whose peoples desire to exercise their liquidity preferences in gold. But, having de-hoarded their gold, they are compelled to store their funds in forms of representative money in which they have little or no confidence. Their panic fears are largely justified. The majority of them have experienced the loss of their wealth in the past by the destruction, or wholesale depreciation, of their national currencies. And now their wealth is held in monetary forms which they regard as equally likely to depreciate.

There is but one antidote to "hot money"—gold. "Hot money" is liquidity preference in a panic. It is in a panic because it is not being exercised in monetary forms having the full confidence of the holder. Fortunately, the national and individual mentalities most susceptible to the panic fears of "hot money" are also extremely susceptible to the soothing powers of gold.

Liquidity preference, or the desire to hold ready money, is an essential and an invaluable factor in all administrative and commercial operations. Without its free exercise, commerce and industry stagnate. But when it is in a panic, when it takes the forms of an irrational suspicion of all investments, or of a mass movement to sell securities, or of a distrust of the forms of money in which it is exercised, or of a mass preference for foreign currencies or foreign securities, it is no longer a valuable financial asset, it is a dangerous liability.

The legitimate exercise of liquidity preference is being denied, and panic forms of money are being created, by such measures as blocked or barter currencies, exchange restrictions, criminal and administrative obstacles to the possession, transfer and export of gold. One of the principal objectives of all monetary administrations should be the provision of adequate quantities of those forms of money in which the liquidity preferences of the owners of funds can most satisfactorily be exercised. If, as is the case in many countries, gold is the popular medium for the exercise of liquidity preference, then the problem of securing an equitable distribution of the large supplies of gold now being produced to all countries in need of its services, is one of the greatest urgency.

CHAPTER V

THE LIMITATIONS OF INVESTMENT

"And I was afraid, and went and hid thy talent in the earth.

"His lord answered and said unto him:

"Thou wicked and slothful servant, thou knewest that I reap where I sowed not, and gather where I have not strawed. Thou oughtest therefore to have put my money to the exchangers, and then at my coming I should have reaped mine own with usury. . . .

"Cast ye the unprofitable servant into outer darkness: there shall be weeping and gnashing of teeth."—Gospel of St. Matthew, chapter 25.

How true to type! On the one hand, the confident city magnate, blatantly boasting that he reaped where he sowed not, instinctively knowing when to buy for a rise or to sell for a slump, an adept in the arts (to use Mr. Keynes' expressions) of beating the gun, of outwitting the crowd, of passing the bad, or depreciating, investment to the other fellow.

On the other hand, the diffident servant, suffering from an inferiority complex, knowing by past experience that if he invested, he put his money into a dud company, or if he ventured to bet, he backed the wrong horse, unable to bear the responsibility of the talent thrust upon him, determined not to lose his master's money, and, finally, trusting to the slogan "safety first," invented for the guidance of men of his limited mentality, burying the money entrusted to his keeping in the earth so that he might restore it, unimpaired, to his master.

The Parable of the Uninvested Talent is a superlative example of the fallacies of pre-War Economics. It regards the storage function of money as a crime to be punished by ejectment into outer darkness and by pains sufficient to cause weeping and gnashing of teeth. The failure to invest the talent entrusted to him was regarded as a reprehensible abuse of money which fittingly merited the epithets wicked and slothful, and the infliction of savage punishment.

But Time has its revenges. Under the imposing designation of Liquidity Preference, the storage function of money has been restored not only to economic recognition, but to economic sanctity. It has been made the basis of new laws, new theories, new conceptions of economic cause and effect. It has relegated the hoary fallacy "Interest is the reward of Abstinence" into well-deserved discredit. Interest is now perceived to be the price paid for obtaining that valuable and potent economic asset, the command over ready money. The storage function of money, from being despised, and ignored, and denied, is now seen to be the vital source from which all payments of interest spring.

The crime for which the wicked and slothful servant was sentenced to be cast into outer darkness has now been firmly established as one of the supreme economic virtues. Governments, Administrations, Institutions, Business Corporations, cannot function unless they follow his example and withhold a considerable proportion of the savings of the community to perform the manifold functions of ready money. Money in liquid form is required in such profusion in the numerous transactions of every advanced community that an accumulation amounting to two-thirds of the total funds of the community has to be stored up to meet the demand for its services.

More remarkable still, the Commercial Banks, the pillars of monetary orthodoxy, instead of inculcating the teaching of the parable and condemning the hoarding of monetary resources, actually force their clients to commit the crime for which the wicked and slothful servant was savagely punished. Under penalty of being forced to pay heavy fines in the form of bank charges, they are compelled to hoard a substantial portion of their current accounts or demand deposits in the form of minimum balances.

And while the storage function of money as a means of accumulating savings has been restored to respectable orthodoxy, the alternative method of providing for the future—investment—has been steadily losing its former prestige. It is now recognised that for the non-expert, investment is a gambling operation, having much in common with the activities of those who patronise the racecourse, the sweepstake, the lottery and the casino.

Investment is not, as classical economics tacitly assumed, the automatic placing of savings in a paying concern producing a certain annual income. It is a highly skilled operation, and if undertaken by a person without the necessary knowledge and skill, it is just as liable to result in a heavy loss as in the gain of a future income. Apart from the operations of a small and selected number of skilled persons, investment is a risky gamble in which the prospect of an income in the future is risked against the future security of the savings invested.

The high moral glamour with which Economic Science surrounded the operation of investment is as false as the deep moral reprobation with which hoarding has been regarded. From a moral point of view, investment is neither better nor worse than hoarding. They are both perfectly natural and legitimate methods of making provision for the future. They have their different appeals to different mentalities and different natures. The skilled, the sanguine, the adventurous, the commercially minded, the person saving for an income rather than for future security, will invest. The non-expert, the cautious, the unbusinesslike, the doubtful, the suspicious, the person saving solely for future security, will hoard. Both are equally justifiable if regarded as a question of abstract ethics.

Both forms of storing savings—hoarding and investment—are also equally justifiable from a purely business point of view. Despite the teaching of imaginary economics, the public is now aware that investors are just as liable as hoarders to be cast into outer darkness and to be compelled to relieve their feelings with weeping and gnashing of teeth.

The first official justification of the growing doubts in the minds of the public as to the superiority of investment over hoarding was afforded by the publication of the Cunliffe Interim Report of 1918. That Report clearly demonstrated that investment was not only a risky form of gambling, but was gambling with loaded dice, with the odds artificially and deliberately rigged against the investor. Should the gold reserves of the Bank of England need to be secured against an external drain, the Bank of England announced that it would adopt a policy of deliberately reducing the value of all investments, and the incomes received therefrom.

The methods by which it was prepared to reduce, and actually did reduce, the value of investments were detailed with great precision in the Report specified above, in the following sequence:

1. the postponement of new enterprises;
2. the reduction of the demand for constructional material and other capital goods;
3. the slackening of employment;
4. the reduction of the demand for consumption goods;
5. the increase in the charges for working capital;
6. the imposition of difficulties in maintaining the supply of working capital needed for carrying on existing enterprises;
7. the artificial causation of falling prices;
8. the forced sale of goods on a weak market.

All these forms of loading the dice and enhancing the risks against the investor were utilised continuously from 1925 to 1931 in order to maintain the obsolete system of currency known as the gold standard.

Thus, to the normal gambling risks associated with investment were added the periodic interventions of the Bank of England, deliberately designed to curtail present investment by making all past investments less remunerative. Investment in any country on a gold standard was not a fair gamble in which the investor risked his savings on his estimate of the future course of industrial development. It was a monstrously unfair gamble in which the investor's estimate of future prospects was liable to periodic adverse and disastrous interventions by the monetary authorities.

It was, however, the Macmillan Report, published in June 1931, which brought home most clearly to the mind of the public the hazardous nature of investment as a method of making provision for the future. This Report, though published six years ago, still remains the most searching criticism of post-War monetary and industrial conditions that has yet been made. It was much concerned with the grave mistrust of investment manifested by the public of all countries since the War. Among its pronouncements on the subject are the following:

> In the modern world where a willingness to assume risk is an essential element of economic life, the process of investment is seriously affected by the reduction of profits due to a disparity between prices and costs of production. An inevitable accompaniment of sharply falling profits is a decline in the yield of ordinary share capital, and therefore a fall in its market price, as has indeed taken place. Under

these conditions it becomes increasingly difficult to finance new enterprise by the issue of ordinary shares.

And:

> So long as the economic outlook is so very uncertain, and the public are disinclined to venture their savings, the additional savings possible to certain classes when prices fall may merely be hoarded for the time being.

Also:

> The perils of War and the post-War period, and the unfortunate currency experiences of many countries have led to a greatly increased preference for liquid assets, as compared with long-term securities.

One of the distinctions of the Macmillan Report was this anticipation of Mr. Keynes' conception of Liquidity Preference in the phrase "greatly increased preference for liquid assets." It admitted as natural and legitimate the inclination of the public to hoard their savings rather than to accept the heavy risks inherent in investment. Instead of condemning hoarding as deserving of banishment into outer darkness, it recognised it as the natural resource of the cautious and the prudent person, unwilling to subject his life's savings to the risks and uncertainties incurred by investing them.

But the most brilliant analysis of the incalculable risks of investing savings in order to provide security for the future yet presented, is contained in the chapter on "The State of Long-Term Expectation" in Mr. J. M. Keynes' latest work on *The General Theory of*

Employment, Interest and Money. The following extracts are given in the hope that they will stimulate recourse to the original:

> Our knowledge of the factors which govern the yield of investment some years hence is usually very slight and often negligible.
>
> If we exclude the exploitation of natural resources and monopolies, it is probable that the actual average results of investments, even during periods of progress and prosperity, have disappointed the hopes which prompted them.
>
> If human nature felt no temptation to take a chance, no satisfaction, profit apart, in constructing a railway, a mine or a farm, there might not be much investment merely as a matter of cold calculation.
>
> It is its precariousness which creates no small part of our contemporary problem of securing sufficient investment.
>
> The element of real knowledge in the valuation of investments by those who own them, or contemplate purchasing them, has seriously declined.
>
> They (expert professionals) are concerned, not with what an investment is really worth to a man who buys it for keeps, but with what the market will value it at, under the influence of mass psychology, three months or a year hence.
>
> The actual private object of the most skilled investment to-day is "to beat the gun," as Americans so well express it, to outwit the crowd, and to pass the bad, or depreciating, half-crown on to the other fellow.

Finally:

> If the animal spirits are dimmed and the spontaneous optimism falters, leaving us nothing to depend on but a mathematical expectation, enterprise will fade and die.

That these are not merely theoretical opinions is proved by the plight of investors in the United States between 1929 and 1933. The highest monetary authority in the land, the Chairman of the Federal Reserve Board, Governor Eccles, testified to the Banking and Currency Commission of the House of Representatives in 1935 that the income of the people of the United States fell from 83 billion dollars to 47 billion dollars in the four years 1929 to 1933. That fall in income, amounting to 36 billion dollars, was shared by speculators and workers, as well as by investors. But it was primarily caused by a fall in income from investments.

During those four years, it was the investors, not the hoarders, who were weeping and gnashing their teeth. The people of the United States were so convinced by this experience of the superiority of hoarding over investment as the best method of storing their savings, that they commenced hoarding gold to such an extent that the banking system of the country collapsed, and the hoarding of gold had to be checked and controlled by criminal legislation.

Investors in Britain between 1929 and 1932, and investors in France from 1931 to date, suffered sufficient losses to cause them to question the vaunted superiority of investment over hoarding, so automatically assumed by economists. Though their experiences were not so disastrous as those of investors in the United States, they were sufficient to make them distrustful of the theory that investment should be the sole and only method of storing their savings and making provision for the future.

The alleged superiority of investment over hoarding

as a store for savings was nothing more than a figment of the economic imagination. Instead of providing security for the future, investment produces illusions, doubts and uncertainties. It is frequently based on slight and negligible knowledge; it seldom realises expectations; as a matter of cold calculation it is not justified; it is largely precarious; it is the result of "animal spirits," "spontaneous optimism," "the temptation to take a chance," rather than of "mathematical expectation." Moreover, investors are at the mercy of skilled professionals, whose object is to beat the gun, to outwit the individual saver, and to pass on to him, as the fruit of his life's savings, the bad or depreciating investment.

There is, therefore, no ground for the opinion that with the spread of education and enlightenment, the post-War movement away from investment and towards hoarding will cease. Economic Science has lost its certainty as to both the moral and material superiority of investment over hoarding for the storage of savings. The investor is just as likely to lose as much by the depreciation of his capital as he is likely to receive in interest. Education and enlightenment, so far from turning the weight of opinion towards investment, is likely to subject investment to a colder and clearer scrutiny as to its ultimate benefits to the individual saver than it has received in the past.

Clear statistical evidence of the marked decline of investment in popular estimation is given in the chapter on "The Accumulation of Capital" in Colin

Clark's recent work on *National Income and Outlay*. After a carefully prepared Statistical Table of British Capital Investment for the years 1907, 1924 and 1929 to 1935, he makes the following comment:

> The course of these figures is very interesting. Investment as a percentage of the national income was as follows:
>
> | | | |
> |---|---|---|
> | 1907 | .. | 12·2 |
> | 1924 | .. | 8·1 |
> | 1929 | .. | 7·2 |
> | 1935 | .. | 6·9 |

Even during a period of rapid recovery of prosperity in 1935, at a time when one form of investment—house building—was achieving record heights, investment as a percentage of the national income was little more than one-half of the pre-War percentage in the year 1907.

The low percentage of investment in the year 1935 was not due to any shortage of savings. After making a comparison of the records of investment with those of savings, Colin Clark states:

> The outstanding fact of the situation is that three channels of savings, namely:
>
> the obligatory saving funds of local authorities,
> trading profits held back by company officials,
> and savings for security by the working and middle classes,
>
> provided enough to meet nearly the whole of investment requirements in recent years, and more than enough at the present time.

The reluctance of persons saving to place their funds

into investments can be illustrated by the figures for the year 1935, as follows:

Savings:	
Undistributed Profits ..	£228 millions
Working- and Middle-Class savings	£172 millions
Government and Administrative savings	£83 millions
Total Savings	£483 millions
Trading losses	£30 millions
Net savings	£453 millions
Investments	£305 millions

Thus, the decline of investment from a pre-War proportion of 12·2 per cent of the national income in 1907 to a post-War proportion of 6·9 per cent of the national income in 1935, is not due to any shortage of savings. There were savings enough, and to spare, in 1935, for a much larger volume of investment.

These facts and opinions have been the commonplaces of the World's Stock Exchanges and Money Markets for some years. They are now rapidly permeating Economic Science. Perhaps the most convincing demonstration of the rehabilitation of the Storage Function of Money, or Hoarding, in the Economic World is given by Professor Robbins, a pillar of economic orthodoxy. In his contribution to *Lloyds Bank Monthly Review* for May 1937, when discussing the question of Unemployed Relief Funds, he stated:

> The root of the difficulty, as we have seen, lies not in the accumulation of a reserve, but in the method of investing and realising it. If instead of *investing* the surplus it accrued, it was *hoarded* in some appropriate form, the effect would be quite different. . . . . But is not this the type of influence for which we have been searching in this article?

The italics in the above passage, emphasising the superiority of hoarding to investment in this particular case of accumulated savings, are supplied by Professor Robbins himself. They indicate that hoarding can be superior to investment in the solution of some of the problems of modern industry.

The antithesis between hoarding and investment is not so great as is popularly imagined. Many modern forms of money, in which savings are stored, have affinities with investment in earning a slight amount of interest; and most investments have a liquidity premium, an attribute which they share with money. There is in fact a continuous chain of financial expedients in modern industry, which bridge the wide gap between the extremes of crude hoarding of gold, on the one hand, and the long-term investment, on the other, by a series of almost imperceptible gradations. From one extremity of this series to the other, each expedient has it own special function to perform. Each form of storing savings supplies some need, provides its own distinctive and special utility, and appeals to some particular type of liquidity preference.

With the growing wealth of the world there will be ample scope for each and all methods of storing savings extending in a series of facilities ranging from the hoarding of gold to long-term investment. The risks of investment will be eliminated for many by such expedients as Insurance, Annuities and Savings Banks. But these methods simply transform individual forms of hoarding into corporate forms of liquidity preference. With every increase in investment there is an insistent growing demand for corporate liquid stores of value with which to finance the large volume of cash payments continuously required in modern industry and finance. These liquid stores are held in various forms of representative money: bank notes, deposits, short-term money. But all such forms of representative money need some material guarantee of their ultimate liquidity, of the maintenance of their purchasing power throughout all countries of the world, and throughout future time.

The economic glorification of investment and the economic depreciation of hoarding have both been due to the neglect of the consideration of the operation of money by economists. The functions of money having been regarded as being symbolic only, the necessity of providing a supply of money of all the various kinds required, equal to the practical material demand for its services, including storage, has never been appreciated by economic science. Nor has the practical administration of money been equal to the provision of its supply in sufficient volume and variety in accordance with that demand. Alternations of periods of too much money, or Inflation, with periods of too little money, or Deflation, and the disastrous wanderings of "Hot Money" from country to country, have been accepted with philosophic resignation as if they were the results of the Laws of Nature.

Inflation, Deflation, "Hot Money," are monetary diseases which have been permitted to attain disastrous proportions because of the neglect of Economic Science to accord to money, and particularly to the storage function of money, the importance it exercises in the actual operations of industry and commerce. They are diseases due to the inability of the monetary authorities to provide either the right quantity, or the right kind, of money required to satisfy the demands of the members of the public in the legitimate satisfaction of their liquidity preferences.

The monetary policies of both the British and American Governments are now being consciously administered with the view to smoothing down the alternations of Inflation and Deflation and thereby eliminating the disastrous operation of the Business Cycle. That can be achieved only by providing the correct volume and the correct kinds of money, and by stimulating its circulation by low rates of interest, in order that it may meet the requirements of the public for its services both as a store of value and as a medium of exchange.

Hoarding fell into disrepute because there has seldom been sufficient money and sufficient varieties of money to permit its free operation. The stringency due to insufficient supplies of the various kinds of money was wrongly attributed to hoarding. The storage function of money is now returning to favourable recognition, partly because it is appreciated that it is a necessary and important function of money, partly because it is now perceived that investment, as a method of storing savings, has not a universal appeal, and partly because enlightened governments can now manage money so efficiently that they can provide sufficient money for all demands for its services.

But a sufficiency of money is not all that is required. It must be provided in the forms that satisfy the desires of the people. Otherwise a portion of it becomes "hot money." The forms of money that appeal to British or American institutions and corporations for the purposes of storage do not appeal to the French peasant, the Eastern European, the South American peon or the Indian cultivator. More material and less representative forms of money are required for individual hoarding than for the commercial storage of funds.

The monetary authorities of most countries have not the capacity to manage their currencies as efficiently as the pound or the dollar are managed. And even if they had such capacity, their peoples do not want their money in the forms in which pounds and dollars are provided. They want metallic money, the more gold the better. And if they are forced against their sounder instincts to use paper money for the storage of their funds, they want it backed by the highest proportion of gold that the monetary authorities can accumulate; otherwise it becomes "hot money" in their hands.

The knowledge that investment can neither fully nor efficiently meet the demand for provision for the future imposes upon all monetary authorities the responsibility of placing at the disposal of the public adequate supplies of money in forms capable of satisfying the liquidity preferences of all. That responsibility has never yet been openly acknowledged. Its necessity has only recently been discovered. But it is significant that on the day the writing of this chapter is being concluded (July 9, 1937) the return of the Bank of England shows the highest note circulation on record (£494 millions). That record circulation is largely due to the heavy French demand for the hoarding of sterling pound notes.

If the Bank of England accepts the responsibility of providing an adequate supply of money in a form acceptable to French hoarders, it cannot be long before the monetary authorities of all countries acknowledge the responsibility of providing for their own people the money necessary to enable them to exercise their liquidity preferences in the forms in which they need to embody them.

That responsibility will necessitate a more equitable distribution of the gold supplies of the world. Fortunately, the current production of gold is sufficient to enable the mal-distribution of gold caused by the enforced de-hoarding of the past fifteen years to be rectified. Before considering the measures necessary to ensure the better distribution of gold, a brief glance at the current supply, and a bold guess at the prospective supply of gold is necessary.

CHAPTER VI

THE SUPPLY OF GOLD

"The stocks of monetary gold which had been slowly accumulated over centuries were virtually doubled in value in five years; while the monetary gold in the United States alone amounted to over 12 billion dollars, a sum equal to the total world supply of 1932. Such a rate of growth in our banking and currency reserves is without precedent in history."—The Hon. R. H. BRAND, Special article in *The Times*, June 16, 1937.

THIS chapter will be mercifully short. The fate of the forecasts of the Macmillan Report as to the future supply of gold, so speedily falsified in the event, affords a warning against any repetition of taking a similar course. But the outlook for gold cannot be effectively considered without a statement showing the amount of recent production and a consideration of the factors tending to affect the amount of the current and future outturn.

The age-old race between the demand for and the supply of gold has had many vicissitudes. One of the most interesting departures from normal was the outflow of the precious metals from the hoards of Mexico, Peru and Brazil following the discovery of America. So great was the increase in the monetary supplies of Western Europe that the precious metals lost two-thirds of their value in the century elapsing between 1550 and 1650. In other words, the price level of commodities was multiplied threefold. Gold, however, survived this shock to its value as a precious metal, a much greater shock than the present large production of gold is likely to cause.

Nothing in the long history of gold is so remarkable as the largely unintended cornering of three-fourths of the world's monetary supplies of gold by the United States and France in the decade 1922–32. These countries attempted, after the war, to copy the gold standard invented by Britain, and controlled by Britain until the outbreak of war in 1914. But, being Protectionist countries, they did not appreciate the fact that the British pre-War gold standard was the basis of a Free Trade international currency. Their gold standards consequently were conceived and administered as intensely national systems of currency, and had no affinity with, nor resemblance to, the British pre-War gold standard.

From the year 1922, the gold standard administration of the United States commenced to drain the gold reserves of the debtor countries. This was the effect, but not the intention of the policy of dollar stabilisation. In 1928, France followed the example of the United States, and the enforced de-hoarding of gold from other countries was accelerated. In particular, de-hoarding was ultimately forced on Britain, and the gold reserve of the Bank of England was reduced to its last sovereign in 1931. Britain was, in effect, abandoned by the gold standard.

Britain's loss of gold was due to her effort to keep

the price of gold down to a fixed level, while the artificial scarcity in the remaining countries of the world, caused by the American and French accumulations, was forcing its price upwards. With the departure of Britain from the gold standard and the cessation of the British effort to keep its price fixed, the Law of Supply and Demand produced its inevitable result. The international supply of gold was so depleted by the national accumulations of the United States and France that it was inadequate to meet the international demand for its services. The price of gold accordingly rose rapidly from 85 to 147 shillings per ounce in Britain, and, subsequently, from 20·67 to 35 dollars an ounce in the United States.

Abandoned by gold in 1931, Britain maintained her international commerce by the brilliant invention of the Exchange Equalisation Fund. But though the gold standard had collapsed, the British Treasury did not make the mistake of deeming that gold had lost its monetary qualities and was no longer able to perform monetary services. As the one and only substance in which the liquidity preferences of all the peoples of the world can be exercised it retains its outstanding monetary importance. As the only universal form of ready money, gold provides the only material link between the present and the future values, and the national and international values, of all forms of representative money. Despite the collapse of the gold standard, gold still remained the only material link between the numerous national currencies of limited circulations and of limited ranges of purchasing power. The operations of the Exchange Equalisation Fund necessitated transactions in gold on an unprecedented scale.

Formed to steady the fluctuations of the international exchanges, the British Exchange Equalisation Fund was soon utilised to perform numerous additional monetary functions, national and international. Among these extra functions were:

1. the protection of Britain from the repercussions of international financial dislocations, in particular, from the drain of its gold supplies;
2. the financing of the international balances of all the nations of the Sterling Group;
3. the protection of the nations of the Sterling Group from external financial dislocations, including the protection of their gold supplies;
4. the re-adjustment of the world wholesale price level to the greatly enhanced price of gold.

These functions required a far greater accumulation of gold than was necessary under the gold standard. Britain's gold reserves mounted, as related in Chapter III, from £150 millions in 1930 to £700 millions in March 1937.

The creation of the British Exchange Equalisation Fund after the collapse of the gold standard, therefore, added a third demand for gold in addition to the two heavy demands of the United States and France under their respective gold standards. They were accumulating gold as the result of their national monetary policies. Britain commenced accumulating gold in the

Exchange Equalisation Fund, for the purpose of restoring the international currency, and to enable her to settle not only her own international balance of trade, but the trade balances of the nations belonging to the Sterling Group. The collapse of the gold standard in 1931 did not diminish the demand for gold. It increased it. And it not only increased the demand for gold, it gave Britain the financial resources in the Exchange Equalisation Fund (£350 millions sterling) with which to finance her enhanced accumulation of gold. Britain set herself first of all to restore the depleted gold reserve of the Bank of England, which had been reduced to zero in 1931, then to make the Exchange Equalisation Fund operate as a cushion to protect her industries and those of the nations of the Sterling Group from the adverse effects of external financial pressure. This entailed the accumulation of large stores of gold in the Fund.

It is now possible to appreciate the forces of demand which stimulated an increase in the supply of gold that, in the words of the quotation at the head of this chapter, virtually doubled the stocks of monetary gold in value in five years, a rate of growth without precedent in history.

The three leading industrial nations in the world, the United States, Britain and France, were in 1932 demanding and accumulating gold to the full extent of their financial resources. This triple demand was far greater than the available supplies of gold. It intensified the de-hoarding that had taken place under the gold standard. India and China were added to the nations forced to de-hoard their supplies of gold.

But the demand for gold after 1931 was of a different nature from the demand prior to that date. Prior to 1931 gold was at a fixed price in all the leading international currencies. There was no rapidly rising price to stimulate its production. After 1931, although the dollar price and the franc price of gold remained fixed, the sterling price of gold was determined, not by the rules of the gold standard, but by the Law of Supply and Demand. As Demand exceeded Supply the price went up. It rose until it forced the United States to raise its fixed price of gold from 20·67 to 35 dollars an ounce. It rose until it forced the countries of the Gold Bloc either to raise their currency prices of gold or to transform their accumulations of gold into de-hoarding.

The rise in the price of gold, to a level between 60 and 70 per cent above its gold standard price, not only stimulated de-hoarding, it stimulated the production of gold. Sources of supply which would not pay when gold was 85 shillings an ounce, became gold mines, literally, when its price was raised above 140 shillings per ounce.

Assuming a uniform valuation of £7 per ounce, the production of gold in the past five years has been as stated in the table on page 97.

The annual outturn of new gold has increased by 45 per cent since 1932. This amounts to an average annual increase of 10 per cent. It is generally accepted that the average annual increase of the production of wealth before the War was 3 per cent per annum, and that general stability was enjoyed because the rate of increase in the production of gold did not depart materially from the rate of increase in the production of wealth.

It is now generally assumed that an annual increase of 10 per cent in the production of gold is far more than

YEAR.	NUMBER OF OUNCES.	£ STERLING.
1932	24,306,000	170,142,000
1933	25,503,000	178,521,000
1934	27,630,000	193,410,000
1935	30,991,000	216,937,000
1936	35,254,000	246,778,000
TOTAL	143,684,000	1,005,788,000

is necessary to finance an annual increase of 3 per cent in the production of wealth. In other words, it is currently believed that the supply of gold promises to exceed demand, and that if so, the price of gold will fall to some new level.

There are several fallacies in this argument both on the side of supply and on the side of demand. The production of gold has been stimulated during the past few years by three factors:

the high price of gold,
the rising price of gold,
the discrepancy between the price of gold and the general levels of the prices of other commodities.

It is the policy of the British and American Monetary Authorities to maintain the price of gold at, or near, its present high level, so the first factor will, as far as can be foreseen, continue to operate.

But the second factor no longer operates. For the future, the production of gold will lose the stimulus of rising prices of gold and will have to adjust itself to a stationary, and not to a rising price level.

The discrepancy between the price of gold and the prices of other commodities is also rapidly being eliminated. This will result in rising costs of production of gold and a lessened stimulus to increased outturn. There is a reasonable doubt whether the annual production of gold will continue to increase at the high rate of 10 per cent per annum when two of the three factors stimulating the increased production for the past five years have ceased to operate. It will probably rise gradually to the production of £300 millions per annum in a few years' time.

But the international gold market (principally Britain and the United States) has been absorbing gold for the past four years at a much greater pace than is indicated by increased production or by an increased supply of 10 per cent per annum. In Chapter III it was shown that de-hoarded gold was being thrown on the international gold market at the rate of £223 millions per annum. The actual supplies of gold for the four years, 1933–36, which have been absorbed have been:

New Gold	£835,646,000
De-hoarded Gold	£891,000,000
Total	£1,726,646,000

It was not the supplies of new gold which caused the gold panic of June 1937. It was the de-hoarding of £70 millions of gold from the French Exchange Fund in that month that gave the momentary impression that the supply of gold was so excessive that the price of gold could not be maintained. The task of absorbing the supplies of new gold has presented no impossible problem to the monetary authorities of Britain and the United States. It was the de-hoarding of a greater quantity of gold than was actually being produced in the years 1933–36, that presented the almost insolvable problem. It created a flood of "hot money" seeking refuge in the British and American currencies. At the same time, it forced those currencies to absorb gold (newly mined and de-hoarded gold) at an average rate of £431 millions per annum.

The problem of dealing with the future supply of gold is not merely a question of new supplies. It cannot be dissociated from the question of de-hoarding. The problem of the past four years has been the problem of dealing with an average supply of £431 millions per annum (£209 millions, new gold, plus £222 millions, de-hoarded gold). The future of gold cannot be estimated by ignoring the second and the largest of these two constituents.

It is necessary at this juncture to resume the consideration of the question of the de-hoarding of gold which was the theme of Chapter III. Is the de-hoarding of gold likely to continue at the same pace as for the last four years? Is the rate of the de-hoarding of gold diminishing, and, if so, is it diminishing slowly or rapidly? Is it likely to cease in the near future? Is it likely to be reversed in direction, and to be transformed into a gradual absorption of gold by the great majority of nations instead of being a heavy drain on their gold resources?

In short, assuming the annual production of gold to rise to £300 millions, will the international gold market have to absorb £300 millions *plus* a large and indefinite amount of de-hoarded gold, or £300 millions *minus* the amount necessary to restore the national currencies of the world to reasonable efficiency?

CHAPTER VII

THE DECLINE OF DISRUPTIVE FORCES

"So long as people have confidence they are not going to call for redemption, but when confidence flees and people begin to get frightened, and realise that we have not got gold with which to redeem, they rush for redemption, as happened when we went off the gold standard. Confidence was shaken. If we had a sufficient metallic base to redeem, confidence would not flee, and we would need only that to hold confidence. Whenever confidence disappears the crash comes. But as I have said, if we had sufficient redemption metal to take care of everybody, and everybody knew it, fear would not come and the crash would not come."—Mr. O. H. Cross, Member of the U.S. House of Representatives for Texas.

The description given by Mr. Cross of the monetary psychology of the people of the United States, applies almost universally to the people of most of the countries of the world.

The people of Britain, having a different currency mentality, find it difficult to appreciate the important monetary qualities of gold to the peoples of many countries in creating confidence, in eliminating fear, in averting panic, and in providing an assurance against financial disaster. Even the experience of the depletion of the gold reserve of the Bank of England to the last ounce owing to the flight of "hot money" from the country in 1931, has not convinced them that, though the people of Britain are largely free from the metallic complex with reference to their money, it is a powerful monetary influence in most other countries.

The coining of the term "hot money" in the United States is evidence of the susceptibility of the American people to the danger of large liquid funds being held by people who have no confidence in the kinds of money in their possession. The danger is all the greater when the "hot money" is of foreign origin. It exists now in far greater volume than when it helped to cause the panics of 1929, 1931 or 1933. It has been poured in large quantities into Britain and the United States, because the monetary policies and the large gold reserves of these two countries give a certain amount of confidence to its holders. It will, however, remain a menace and a danger to the monetary stability of these two countries until the enforced de-hoarding of gold ceases, and each country can restore its gold reserves to the volume necessary to provide all the various forms of money needed for its own monetary requirements.

For the full and free exercise of its industrial and commercial activities, each and every country in the world requires a sufficiency of gold to satisfy the following requirements:

1. to enable each individual citizen to satisfy his liquidity preference, that is, his desire to hold ready money, in gold, if he so desires;
2. to guarantee the adequacy of the amount of each and

every kind of representative money required to finance its industry and commerce;

3. to guarantee the liquidity and future purchasing power of all forms of national representative money in storage or in circulation;
4. to finance any adverse balance of trade on its international transactions;
5. to finance its obligations to pay interest on, and to provide sinking funds, if necessary, for, all loans obtained from other countries, in so far as these cannot be financed by exports of produce;
6. to finance the purchase of foreign currencies and foreign securities to the full extent its citizens desire to hold them;
7. to finance any foreign loans its citizens may be in a position to make to other countries.

Some countries want large stores of gold, some less, for these purposes. Some currencies are used for internal and national purposes only; others perform world-wide international functions. Some peoples have a highly developed metallic complex, or gold complex, with respect to their currencies; others are entirely free from such complexes. Some exercise their liquidity preferences mainly in gold, or in gold only; others are satisfied to exercise them in deposits and recognised forms of short-term money. In some countries, the majority hoard their savings; in other countries, the majority invest them. Some are debtor countries; others are creditor countries. Some try to be self-sufficient; others are dependent upon international commerce for their existence.

All these divergences are reflected in the amount of the store of gold necessary for the full industrial and commercial development for each country respectively. It is doubtful if any country, even the United States or Britain, has sufficient gold for all these requirements. Britain has been compelled to supervise and restrict her foreign loans to other countries in order to safeguard her gold reserves. The United States not only prohibits loans to certain countries, but also denies its citizens the right to exercise their liquidity preference in the way they desire, by limiting, almost to the verge of prohibition, the possession and export of gold.

The majority of countries do not possess sufficient gold to satisfy any one of these sevenfold requirements. Their peoples have been compelled, by the forced de-hoarding of their national stores of gold, to exercise their liquidity preference either in bank notes or in forms of "hot money," neither of which possess their confidence. Their stocks of representative money are neither adequate in volume nor sufficiently liquid to purchase the requirements of their holders. Some countries are compelled to adopt obstructive restrictions on international trade because, owing to inadequate gold reserves, they cannot finance their imports from other countries. Others have defaulted on their international obligations, for the same reason, inadequate gold reserves. Most countries impose every possible obstacle to the purchase of foreign securities or currencies. Indeed, the monetary and commercial freedom of the peoples of many countries is limited, and their industrial efficiency and prosperity are severely handicapped in all directions by the inefficiency and insufficiency of their national currencies, due to the inadequacy of their gold reserves.

The rehabilitation of their currencies requires the restoration of the gold drained from their resources during the last fifteen years, plus an addition to ensure future progress. Some measures already taken have tended to stop the drain of gold and to secure a more equitable distribution of world gold supplies. The first step in this direction was the formation of the Sterling Group of nations. By linking their national currencies to Sterling instead of to gold, and by placing a portion of their gold reserves in London, these nations protected themselves from the drain of gold from which most other nations continued to suffer.

The next step taken to ensure a more equitable distribution of gold was the adoption of a Commodity Standard of Currency by Britain at the Ottawa Conference in 1932. The first sentence of the Ottawa Monetary Report, in which this momentous currency revolution was announced, reads as follows:

> A rise throughout the world in the general levels of wholesale prices is in the highest degree desirable.

A rise in the world wholesale price level was precisely what was needed to reduce, and ultimately to stop, the de-hoarding of gold from the countries producing primary commodities. The low prices caused by the deflation of the international currency produced by the conflicting gold standards operating from 1925 to

1931, had prevented those countries from selling their produce at remunerative prices. The world wholesale price level fell by 48 per cent between 1925 and 1932. Unable to square their balances of trade by the export of their produce, they were compelled to settle their adverse international balances by gold payments. The adoption by Britain of a currency policy having as its primary objective the raising of the world wholesale price level is now operating, first to reduce, and then to stop the de-hoarding of gold, and ultimately to enable the countries which have suffered from a drain of gold to replenish their depleted stores of gold.

The improvement already effected in this respect in Indian conditions is discussed in a special article on "India and her Gold" in the *Economist* of June 19, 1937. The trend of events is therein expressed in the following query:

> With growing gold supplies outside, and with India's expanding trade balance, is it not now possible that the gold flow will become inward rather than outward?

The Monthly Review of Barclays Bank for August 1937 associates the progressively declining export of gold from India with the increase in her export of commodities. The same economic forces which affect India, and which are tending to transform an outward flow into an inward flow of gold, apply to all countries producing primary commodities. The rising prices received for their produce, as the result of British monetary policy since 1932, increases the value of their exports, improves their international trade balances, and confers upon them a greater capacity to import the commodities they require. In so far as they require gold (and their need for gold to establish their currencies on a sound basis is intense) their increasing capacity to import commodities means an increasing capacity to obtain the much-needed commodity, gold. In the words of the passage quoted above, there is a promise that the gold flow will become inward rather than outward.

The British Empire Monetary Declaration of July 27, 1933, not only confirmed the persistence of British monetary policy:

> in furthering the rise in wholesale prices until there is evidence that equilibrium has been re-established;

which had been proclaimed eleven months before at Ottawa, it also defined the equilibrium position as being one:

> which ensures an economic return to the producer of primary commodities;

Also:

> which harmonises the burden of debts and fixed charges with economic capacity.

Thus, British monetary policy is being consciously directed towards easing the financial burdens of the debtor countries and the countries producing primary commodities, thereby saving them from the forced de-hoarding of their gold, and ultimately providing them with the resources enabling them to purchase gold in accordance with their requirements.

The same objective is also presented in the following passage of the same Declaration:

> the deliberate management of the international standard in such a manner as to ensure the smooth and efficient working of international trade and finance.

"The smooth and efficient working of international trade and finance" is another way of expressing the elimination of the disastrous method of financing international commerce by the disruption of national currencies through the forced de-hoarding of gold.

The working of the Exchange Equalisation Fund, by providing a revolving fund of £350 millions (now £550 millions) for international currency, is the principal executive measure adopted to carry out the declared British policy of raising the world wholesale price level and of ensuring the smooth and efficient working of international trade and finance. Although the accumulation of gold in the Fund increased de-hoarding for a time, it has in its general operation been a potent force in diminishing and stopping de-hoarding from the debtor countries and from the countries producing primary commodities, by raising the prices of their produce.

But while British monetary policy was providing a sufficiency of international currency to raise prices and effect a restoration of the conditions leading towards a more equitable distribution of gold, a contrary policy was being followed by the countries of the Gold Bloc. By remaining on a particularly deflative form of gold standard, they permitted their currencies to become over-valued in comparison with the pound and the dollar. This neutralised to a great extent the British policy tending towards a more equitable distribution of gold. The over-valuation of their currencies caused their exports to decrease and forced them to settle their adverse international balances of trade by gold payments. It also caused the export of capital to countries with lower valued currencies, the export of capital being financed by an export of gold. The de-hoarding of gold by France, Germany, Italy, Holland and Switzerland between the years 1933 and 1936, as described in Chapter III, was due to the attempts of those countries to remain on an impossible gold standard.

This disastrous stimulus to the de-hoarding of gold was stopped, to a certain extent, by the Three-Power Monetary Agreement (subsequently expanded into a Six-Power Monetary Agreement) of September 1936. Unfortunately, under the Agreement, France was not permitted to withdraw completely from the gold standard. The franc was still tied to gold by a fixed minimum gold content below the value of which it was not permitted to fall.

The franc, in comparison with other currencies, being still over-valued, the de-hoarding of gold from

France continued. The French Exchange Fund of £100 millions was gradually depleted by a further flight from the franc, which culminated in the de-hoarding of £70 millions of gold in June 1937. The last link which bound France to the gold standard was then snapped, and the franc was permitted to find its natural level. This is an essential condition for the stoppage of the de-hoarding of gold from France. Other factors, including the balancing of the Budget, are also necessary, before de-hoarding of gold from France can be regarded as being completely checked. Measures to secure this objective are being taken. At the time of writing, the future of the franc is obscure, but it is permissible to assume that the de-hoarding of gold from France will be stopped as a result of the policy being followed.

But though the Three-Power Monetary Agreement did not stop the de-hoarding of gold from France, it succeeded in stopping such de-hoarding from Holland and Switzerland. The four countries, Germany, Holland, Belgium and Switzerland, have not only ceased de-hoarding; they have recently commenced to replenish their stocks of gold, a process which is still continuing at a moderate pace.

The de-hoarding of gold from China and the South American countries has also ceased, thanks to rising prices and improving industrial conditions. Arrangements have already been made by the monetary authorities of the United States to supply gold to China in exchange for silver, and also to Brazil in exchange for produce. Gold is also moving from Britain to Argentina.

Assuming that the steps taken to rehabilitate the franc will succeed in stopping the export of capital and the de-hoarding of gold from France, the only countries which will still be de-hoarding gold are India and Japan. De-hoarding by India will probably cease shortly. If so, Japan will remain the only country de-hoarding gold.

The case of Japan is special, and requires separate explanation. The raising of the world wholesale price level which has stopped the de-hoarding of gold from the countries producing primary commodities has acted in the contrary manner towards Japan. The heavy rise in the price of primary commodities has increased the cost of the Japanese imports required for her heavy armament programme, an extra cost which is at present being met by the de-hoarding of her gold reserves. This is not likely to be continued indefinitely. It will probably cease when Japan has expended the profits obtained by the re-valuation of her gold reserves.[1]

Thus, the de-hoarding of gold has had several successive causes in the past fifteen years. First, the heavy deflation of the international currency between 1925 and 1933 reduced the world wholesale price-level by 48 per cent and forced the countries producing primary commodities to square their international balances by de-hoarding their gold. Secondly, the

[1] This was written before the outbreak of the Sino-Japanese War.

European Gold Bloc by keeping its several national currencies at a high level in comparison with the pound and the dollar stimulated the export of capital and the de-hoarding of gold. And now expenditure on armaments is delaying the restoration of the gold reserves of some countries, their resources obtained from the higher prices of their products being spent on armaments rather than on gold. In one case, Japan, expenditure on armaments has actually caused the de-hoarding of gold.

Briefly, the history of the de-hoarding of gold is that commencing in 1922, it gradually rose to an average of £70 millions per annum in the period 1928–31; then, stimulated by other forces, it rose still further to an average of £223 millions per annum for the period 1933–36; and has now (July 1937) fallen to a small and declining volume from three countries only, France, India and Japan.

Thus, monetary forces are now operating to reduce the amount of de-hoarding that has caused such a mal-distribution of gold in the course of the past fifteen years. In some cases—Germany, Holland, Belgium, Switzerland, Poland, Czechoslovakia, Brazil, Argentina and China—de-hoarding has already been reversed and transformed into a moderate replenishment of gold supplies. The improved efficiency of the international currency, financed as it is by a number of active Exchange Funds, operating in close co-operation with each other, has been largely instrumental in effecting this result. It is, in short:

The deliberate management of the international standard in such a manner as to ensure the smooth and efficient working of international trade and finance,

which forms the objective of the British monetary policy according to the British Empire Currency Declaration of July 1933, and which has now been accepted as the policy of the Six Powers under the Six-Power Monetary Agreement of 1936.

But so far its benefits have been largely negative. The mal-distribution of gold caused by a long period of de-hoarding still persists. Few nations, if any, have sufficient gold to fulfil the seven-fold requirements of a national currency, specified at the commencement of this Chapter. It may be assumed that the Commodity Standard of international currency initiated by Britain in 1932, and tacitly accepted by the leading industrial nations in the Three-Power and Six-Power Monetary Agreements of 1936, will in due course raise world wholesale prices to an equilibrium level.

But even if by the operation of international currency the world wholesale price-level can be stabilised, few nations have the capacity to operate and to stabilise their internal currencies without a substantial backing of gold. The stoppage of the de-hoarding of gold is not sufficient. It must be supplemented by an active distribution of gold to all countries in need of its services. There are ample gold supplies for this purpose. The slow replenishment of their gold stocks by a few nations needs to be accelerated and extended to all nations. The forces which have been effective in stopping the de-hoarding of gold must be supplemented so that the balance between the gold reserves necessary for the efficient working of the international currency and the gold reserves necessary to the efficient working of the various national currencies can be restored.

CHAPTER VIII

THE RE-DISTRIBUTION OF GOLD

"We must have some co-ordination through the central authority that is perpetual between the stock of money in this country, and the pieces of paper and forms of credit which are convertible into money, otherwise you do not have it convertible. We are in a position where, if some event alarmed the public, the drop in price proceeding to an alarming extent, the people were afraid of their bank deposits not being convertible into cash, and rushing to the bank, where they are unable to get more than ten per cent."—Mr. J. H. RAND, Junr., Chairman, Remington-Rand, Inc.

THE above quotation and the quotation at the head of the previous Chapter demonstrate how intimately the monetary system of the United States is dependent on the magnitude of its gold reserves. The question of the re-distribution of gold must be considered with a full appreciation of the fact that the internal monetary system of the United States requires an extremely high accumulation of gold reserves to guarantee the liquidity of the enormous and complex superstructure of representative money in circulation and in storage in that country.

The internal monetary system of France requires an even greater accumulation of gold reserves in proportion to the volume of industry and commerce to be financed, than is necessary for the United States. The heavy accumulations of gold by France and the United

States have been effected in response to a genuine and imperative demand to provide the metallic backing to their national currencies necessary to satisfy the monetary psychology of their peoples.

The remaining countries of Europe (the Scandinavian countries excepted) require an even higher proportion of gold reserves for the satisfactory operation of their currencies than is required by the United States and France. The same holds true for the countries of Asia and South and Central America.

Not only do these countries require large gold reserves for the efficient functioning of their national currencies, their peoples require large supplies of gold coinage for the efficient functioning of their individual liquidity preferences. They have been denied the satisfaction of this traditional and legitimate demand by the withdrawal of gold coinage since the outbreak of war in 1914. It has been incorrectly assumed that because the people of Britain have no use for a gold coinage, therefore all the peoples of the world, with widely divergent monetary mentalities, are equally indifferent to the virtues and the value of gold. The existence of "hot money" demonstrates the magnitude of the danger caused by this misconception.

Perhaps the most effective demonstration of the vitalising effect of adequate gold reserves on a national currency is supplied by Russia. Six years ago the currency of Russia was a by-word and a reproach. It was insufficient. It had neither a constant nor a uniform value. Its circulation and its purchasing power were subject to numerous administrative restrictions which prevented it from being a uniform and stable medium of exchange.

But, thanks to the enhanced production of gold within the country, its currency has been completely reformed. It is not yet adequate for the country's needs, but it is expanding towards that objective. Most of the restrictions on circulation and purchasing power have been removed, and in uniformity of value and stability of purchasing power it is now superior to many of the European national currencies.

The effect of expanding gold reserves on the international commerce of Russia is equally favourable. Russia is no longer compelled to export the wheat needed by her half-starved people in order to pay for the import of machinery and equipment required for her ambitious industrial plans. Her gold supplies are ample to settle her balance of trade. The possession of adequate gold supplies has given her the power to mould her own destinies that was lacking when she was dependent on inadequate gold reserves and inefficient currency management for her monetary resources.

The Russian example exposes the futility of the exhortations addressed to various countries to abandon their quotas, their exchange restrictions and controls and their blocked and barter currencies. It is impossible for them to do so in the absence of adequate national gold reserves. Such restrictions were imposed, in the first instance, to try and stop the devastating de-hoarding of gold to which they were subjected by the fall in the price of their products. As they could not sell their exports at remunerative prices, they were compelled to impose restrictions on their imports; at first in order to attempt to save their gold reserves from depletion, and when their efforts proved unavailing and their gold reserves had vanished, to save themselves from bankruptcy through having no gold with which to settle their balances of trade.

Russia has been able to return to a comparatively efficient monetary policy, internal and international, because her adequate and growing gold reserves have given her the power to settle her balance of trade, and the confidence to restore her national currency. Other countries will be able to abandon their quotas and exchange restrictions, and to dispense with blocked and barter currencies, as, and when, their gold reserves are restored to such an extent that they can discard these artificial monetary defences without jeopardising their national existence. These obstacles to international commerce are the effects of unhealthy internal monetary conditions. They cannot be removed until their internal monetary systems are restored to health by adequate gold reserves.

The equitable distribution of gold in the nineteenth century was achieved by the British administration of the gold standard on international lines. Being then a Free Trade nation, Britain received payment for her favourable trade balance in goods, and not in gold, thereby permitting the debtor nations to acquire and retain the gold necessary for the efficient working of their national currencies. As the Macmillan Report demonstrated, this widespread and beneficial world distribution of gold was disturbed by the national gold standards of the United States and France, which required payment for their favourable trade balances from debtor countries in gold, and not in goods. The gold so acquired at the expense of the debtor countries was genuinely needed by the United States and France for their internal monetary requirements.

The enhanced production of gold in recent years has made the remaining nations of the world independent of the large accumulations of gold acquired by the United States and France. Large gold reserves are required by the monetary authorities of both these nations, who are compelled to base their monetary policies on the currency requirements of their peoples. But there is now enough gold being produced to supply in full not only the requirements of the United States and France, but also the requirements of each and every other country for their national currencies, and for the satisfaction of the liquidity preferences of their peoples, and also, in addition, to provide sufficient gold for the purposes of international finance.

Despite the momentary panic of the spring of 1937, the United States has not too much gold for her monetary requirements. In the first four months of 1937, the currency authorities of the United States were required to finance the following international monetary displacements:

Adverse Trade (and Service) Balance	..	− $132 millions
Long-Term Capital Movement ..	..	+ $792 "
Short-Term Capital Movement ..	..	+ $404 "

The Short-Term Capital Movement is foreign "hot money," liable to be withdrawn at very short notice. Thus, the first four months of 1937 made the monetary authorities of the United States liable to a drain of gold amounting to $536 millions ($132+ $404 millions). It is true that this liability was more than covered by the current gold movement to the States. But should conditions in Europe favour a repatriation of capital, the United States would have to face a drain of gold of considerable dimensions.

Despite the complex internal monetary problems of the United States, and the continuous liability to a drain of gold to Europe, the monetary authorities of that country are already taking the first steps to effect a re-distribution of gold to the countries most in need of it. Already substantial help has been given to Brazil and to China, in the shape of consignments of gold, towards the restoration of the gold reserves of these countries. The United States will probably find a remunerative field for the disposal of any surplus gold that may be acquired in the restoration of the gold reserves of the countries of South America.

Britain is also in a position to take practical action towards effecting the re-distribution of gold. It is one of the few countries in which the purchase, the sale, the possession, and the export of gold are free. As the debtor countries find that, owing to the rising prices of their products, they can afford to purchase gold, there is a perfectly free market in Britain in which they can replenish their gold reserves at market price. Hitherto the price has been too high for them. But now that the world wholesale price-level is rising to an equilibrium with the price of gold, the relatively high price of gold will vanish and will not be a barrier to the purchase of gold.

In order to demonstrate her complete and absolute independence of the rigidities of the gold standard, Britain buys and sells gold sometimes at a discount below the American gold standard parity and sometimes at a premium above that parity. The British price of gold is kept in healthy contact with the Law of Supply and Demand, as far as that Law is allowed to operate, as well as being kept in touch with the movements of the foreign exchanges. Any country with the necessary resources can freely purchase gold for the stabilisation of its currency at the prevailing London market price.

But will this demand for gold—the purchase of gold by countries which have previously suffered from enforced de-hoarding, in order to restore their gold reserves—be sufficient to absorb the large new supplies of gold now being produced? That is a difficult, if not an impossible, question to answer.

All such questions regarding the future are obscured

by international unsettlement. If a period of international tranquillity could be assured, then it is probable that the restoration of the gold reserves of the nations which have been the victims of gold de-hoarding during the past fifteen years would, together with the demand for international currency purposes, absorb all the forthcoming gold supplies for many years to come.

But, as things are, they may prefer to use their greater resources to build up their armaments rather than to build up their gold reserves. The case of Japan affords an illustration of this possibility. The rise in the price of primary commodities which is bringing prosperity to many nations, is adding greatly to the expense of producing armaments. So great is this extra expense of purchasing raw materials, that Japan has been forced to join the de-hoarding nations and export gold in increasing quantities in order to purchase the requirements for her armament policy.

If many nations follow the example of Japan and increase their armaments rather than their gold reserves, it is possible that their combined demands for gold may not be sufficient to absorb the large new supplies of gold.

But the demand for armaments, though at the moment it may be frustrating the re-distribution of gold to the countries most needing its services, will, in the long run, greatly facilitate such re-distribution. The American experience of the non-payability of war debts will limit the sale of armaments and war stores, or even of the fundamental necessities of existence, on credit to the combatants in any future war. One of the most potent forms of armament is a large national reserve of gold. It ensures a supply of armaments and necessaries from neutral nations in times of emergency. Existing armaments rapidly become obsolete. A large store of gold ensures a supply of the most up-to-date war implements that can be purchased.

The present international outlook is dominated by the fear of war. It may seem callous to regard this fear of war as an instrument for securing the re-distribution of gold. But gold is traditionally more influenced by realism than by idealism. As the impossibility of securing armaments and necessaries by credit in time of war is realised, the importance of securing as large a proportion of the newly produced gold as possible will be appreciated. The nations fearing attack will need a substantial war chest of gold even more than the nations which have prepared themselves for offence.

The combination of the policies of cheap money and a free gold market is the British contribution towards effecting the re-distribution of gold. Other things being equal, gold tends to flow to those countries willing to pay high interest rates for its services. More positive measures to assist re-distribution must await the determination of the amount of gold required by Britain for her national and international requirements.

How difficult it is to determine Britain's gold requirements can be appreciated by a brief reference to the

Report of the Macmillan Commission. Discussing the amount of the British note circulation it stated:

> The absolute maximum (note circulation) beyond which the Bank of England must not go might be set at £400 millions, this figure being subject to modification by law from time to time.

As regards Britain's monetary reserves, the Report recommended as follows:

> We suggest therefore that in addition to gold reserves fluctuating between £100 millions and £175 millions, the Bank might maintain other foreign liquid resources fluctuating in amount up to £50 millions.

Six years ago £400 millions was regarded as "the absolute maximum" note circulation only to be exceeded by special legislation. That absolute maximum has now been passed by over £100 millions in the ordinary routine of monetary administration. The reserves considered ample six years ago, ranging from £150 millions to £225 millions, now approach £800 millions, and provision has been made for them to reach £1,000 millions if needed.

In view of these rapidly expanding monetary requirements calling for expanding supplies of gold, it is necessary to make a generous allowance for future requirements of gold. Both British and American monetary policies are directed to the elimination of the business cycle, the restoration of international commerce, and the maintenance of world wholesale prices at a high remunerative equilibrium level. Such policies in the present stage of monetary technique require large and increasing supplies of gold. Much caution must be exercised before it is possible to determine with certainty that either Britain or the United States have redundant supplies of gold.

But it is impossible to present the complete problem of the "Outlook for Gold" without taking into consideration the possibility that both Britain and the United States may, in the future, find themselves absorbing more gold than they can usefully employ. The proposals about to be suggested must not be regarded as being recommended for immediate application; possibly they may not be required at all. They are put forward to demonstrate that, in the event of it being established beyond all manner of doubt that either Britain or the United States should have, in the future, more gold than is needed for their requirements, the problem of re-distributing it to nations insistently needing its services is not insuperable.

Such a position is imagined by Paul Einzig in Chapter XLVI of the *Theory of Forward Exchange* in the following passage:

> It is conceivable that a centre with a low Bank Rate may wish to encourage an efflux, either because the funds are not wanted in the country, or because the authorities want to assist countries which are in need of an influx.

In these circumstances any gold supplies surplus to British requirements can be effectually re-distributed through the Exchange Equalisation Fund. The administrators of the Fund are finding themselves more and more compelled to have recourse to Forward

Exchange operations in securing international equilibrium. By depreciating forward Sterling a little below parity, they can both check the rate of the inflow of gold into Britain and establish an outflow to other countries.

Believing that the United States has actually arrived at a stage where the inflow of gold may need to be checked, Mr. Einzig makes the following suggestion:

> It would not require much effort by the United States authorities to cause an artificial depreciation of forward dollars, thereby checking the influx of artificial funds, and even causing an outflow.

The utilisation of the potent method of Forward Exchange would help to re-distribute gold bullion to those countries from which gold has been de-hoarded in the past fifteen years. The United States and Britain can assist France to replenish her depleted stores of gold by this method.

An even more effective and drastic method would be to effect the re-distribution of gold by retail rather than by wholesale methods, by coinage rather than by bullion. This would attack the problem of "hot money" as well as the problem of the mal-distribution of gold. The epidemic of "hot money" dates from the withdrawal of gold coinage. It cannot be cured except by providing the holders of "hot money" with the kind of money in which they desire to exercise their liquidity preference.

Britain can, if necessary, solve the problem of the over-supply of gold at short notice by restoring the coinage of sovereigns and half-sovereigns, and by issuing them freely to the public through the banks at a price to be notified daily, exactly as the wholesale price of gold is notified daily.

This solution will be necessary only in case the demand for bullion for the replenishment of national gold reserves falls short of the available supply of gold. The restoration of national gold reserves is the most insistent and the most effective measure for the restoration of international commerce and general world prosperity. No other utilisation of gold should take precedence of this first call on its services. Moreover, if national gold reserves could be restored, then the responsibility of providing gold coinage to satisfy the liquidity preferences of individuals would no longer fall on the shoulders of the British monetary authorities.

But if, for one reason or another, the demand for national gold reserves does not materialise, or is too slow, or too small, to absorb the supplies of new gold, then the resumption and issue of gold coinage provides an easy and practicable solution of the problem of the disposal of the surplus supplies of gold. It is far preferable to any alternative solution suggested. The lowering of the price of gold substantially below present levels would cause a monetary deflation that would reproduce the disasters of the years 1929–33.

In a special article on Banking and Trade in *The Times* of July 2, 1937, it is correctly stated that the suggestion that the American price of gold should be lowered came from "academic circles," and not from the industrial and banking and business communities.

The following extracts from that article faithfully present the nature of the panic caused by that academic intrusion into the world of finance, and the disastrous consequences that would have resulted if the American monetary authorities had taken the proposals seriously:

> It was suggested that America, realising that she was the chief purchaser of gold, the main supplies of which come from the British Empire, might reduce her price for gold, say, from $35 to $30 an ounce and thus check production. The mere suggestion of such an action had an immediate and profound effect upon the commodity and security markets, and, indeed, upon the whole business community.
>
> Instinctively producers, whether farmers or manufacturers, felt that any sharp reduction in the price of gold would be followed by a fall of possibly corresponding amount in the commodity price level. The instinct was soundly based since the experience of every country which has devalued or depreciated its currency in terms of gold is that a rise in other commodity prices sooner or later takes place to a roughly corresponding extent.
>
> The economic records of France, Germany, Belgium, and other countries which have depreciated their currencies clearly point to the fact that while governments have the power to alter the currency price of gold they have no power to vary the value of gold in terms of other commodities. In other words the value of gold in terms of other commodities is determined by a number of economic factors, and ultimately there is no artificiality in the relationship in which gold and other commodities exchange for one another in an open market.

Just as the depreciation of a currency causes commodity prices to rise, so the suggested appreciation of the dollar would cause, and actually did cause, prices to fall.

Thus, of the two alternatives:

1. allowing the price of gold to fall, by refusing to acknowledge and to satisfy the insistent demand for gold throughout the world for the satisfaction of individual liquidity preferences;
2. utilising the abundant supply of gold to allow it to perform the individual storage services which originally created its value and which have been freely available to all mankind for fifty centuries up to the outbreak of war in 1914;

it is not difficult to decide which will be the more effective in solving the gold problem.

As to the demand for gold coinage to satisfy the liquidity preferences of individuals, from France alone there would be a demand equal to the forty milliards of francs hoarded in French notes, plus an indefinite sum, estimated between £20 millions and £40 millions, hoarded in British pound notes. The British sovereign has always been a popular coin with hoarders throughout the world. During the nineteenth century there was a constant seepage of gold coins in accordance with Gresham's Law from British circulation for purposes of hoarding, and many were exported for hoarding in other countries. In India, the golden sovereign is the favourite coin for hoarding, and sovereign gold is the favourite quality of gold used for making ornaments.

There is not the slightest doubt that the French holders of franc notes and pound notes would welcome the opportunity to transfer their hoards from paper francs and paper pounds into golden sovereigns if they were given the opportunity. So would the holders of all forms of "hot money" in all countries. The existence of "hot money" is a symptom that its holders have not found a suitable medium in which to store their savings or their liquid funds. They distrust long-term investment in which to place their resources, and they equally distrust all forms of money at present at their disposal. They want gold, and they won't be happy till they get it. The existence of an abundant gold coinage, in which all persons, who wish to do so, could store their funds, would rid the world of the dangerous menace of "hot money."

Such coinage, when issued, would not remain in circulation. Its currency value would change, if not from day to day, from week to week. That does not matter in the least. It is not wanted for circulation. It is required for storage. The public has been educated up to the knowledge that a golden sovereign is not equivalent to a pound sterling. They know that the price of gold is not fixed. They have sold their sovereigns at rates varying from 30 to 35 shillings, and will not be in the least nonplussed by the issue of sovereigns with varying monetary prices, as the wholesale price of gold varies.

Nor will the fact that such gold coinage would have a varying currency value and would not circulate as money have any adverse effect on its storage value. Gold is a precious metal eminently suitable for storage quite independent of its value as a medium of exchange. The Government would probably arrange a buying price for sovereigns at, say, one penny below their selling price for the day. That would, if necessary, maintain their value.

But that would not really be necessary. In India, the traditional home of gold hoarding, gold does not circulate as currency. The circulating medium is silver. Sovereigns are greatly treasured in India, not because they are, or ever were, in circulation, but because they are gold. Added value was given to them because of their beauty, because of their convenience for hoarding, and because sovereigns had exactly the right degree of hardness to provide a most suitable quality of gold for making ornaments.

The necessity for providing gold coins for storage purposes is presented most convincingly in that compendium of the monetary experience of mankind known as Gresham's Law. In its popular form this law states:

> Bad money drives out good money.

Its full meaning is that good money (that is, money with the highest intrinsic value in circulation) is specially selected by hoarders to add to their stores of savings, and consequently passes out of circulation. A fuller statement of Gresham's Law would be:

> Bad money drives good money out of circulation into storage.

The existence of "hot money" in such profusion is a demonstration that modern forms of representative money do not provide the forms of money needed for both circulation and storage. "Hot money" is Greshamic bad money. There being no Greshamic good money in circulation, there is none available for such storage of funds as is in demand by individuals who do not wish to invest them. The individual storage of funds must therefore be effected by forms of money (deposits and short-term money) eminently suitable for circulation and for performing the short-term functions of a medium of exchange, but not, in the opinion of the holders of the funds, suitable for keeping their funds in permanent storage.

"Hot money" is a convincing demonstration that the limitation of the functions of money to two only, those of providing a measure of value and a medium of exchange, advocated by eminent economists, as cited in Chapter IV, is a serious error having disastrous consequences. "Hot money" is money that is not functioning as a measure of value, and is not functioning as a medium of exchange. It is money, devised for the purpose of circulation, that is functioning (very badly) as a store of value. "Hot money" is evidence of the frustration of Gresham's Law. This Law emphasises the necessity for two kinds of money expressed in archaic Elizabethan language as "bad money" and "good money," one effective for circulation only, and one essential for storage. Translated into modern monetary terms these two qualities of money are:

1. Representative money, designed to perform the short-term function of a medium of exchange; and, with an adequate gold backing, the storage function of money for advanced industrial communities;

2. Coinage, designed to perform the long-term function of the storage of individual funds, to be provided by some medium possessing universal acceptability.

These two qualities of money are required not only in order that the storage function of money can be efficiently performed, but that money itself shall maintain its value and its utility in performing all its functions.

The persistence of the demand for gold in providing the valuable monetary service of liquidity is demonstrated by *The Times* Money Reports for the last week of September 1937. The Report for September 24, 1937, commenced as follows:

> Bullion brokers reported an increase in hoarding of gold, and an appreciable proportion of the larger amount of £702,000 dealt in at the "fixing" was believed to have been taken for that purpose. It was reflected in the premium of ½d. over dollar parity.

After daily references to the strength of the hoarding demand for gold, the Report for October 2nd stated:

> The demand for gold showed marked expansion yesterday. At the gold "fixing," £1,070,000 was sold at a price showing an increased premium of 2d. over dollar parity. After the fixing further appreciable amounts of gold were sold, the premium continuing to increase and finally reaching 3d.[1]

Thus, despite the severe repression and the numerous restrictions on the hoarding of gold, it has sufficient vitality to create a substantial premium over exchange parity. The monetary qualities of gold make it an

[1] The premium subsequently rose to 1/5, and for the last three months of 1937 Continental hoarders have been paying a higher price for gold than 35 dollars an ounce.

efficient antidote to the disturbing effects of "hot money."

The strength of the hoarding demand for gold, not only in the form of bullion, but in the form of coinage, is demonstrated, somewhat ironically, by the American Economist, Hans R. L. Cohrssen, in the following passage from his article on "Gold and the Dollar" in the July issue of *Dynamic America*. He states:

> Instead of circulating gold coins, which would be soon snatched up by hoarders, gold certificates could be issued.

In this passage Mr. Cohrssen proclaims, and at the same time repudiates, the perfect solution of the two problems of the over-supply of gold and the menace of "hot money" by the issue of gold coinage.

The existence of a strong and active demand for coinage in addition to the demand for gold bullion is demonstrated by the following passage from *The Times* Money Market Report for October 4, 1937:

> A brisk demand for gold on hoarding account was again reported on the bullion market, sovereigns being also on request at a considerable premium.

The quotation at the head of the first Chapter of this Volume presents the opinion expressed by Mr. J. M. Keynes that when gold is out of sight and no longer passes from hand to hand, it may lose its popular value. That is a very real danger. The value of gold was created originally by the free exercise of the liquidity preferences of individuals, not by the demands or the action of governments. It can only be maintained by the popular acclaim by which it was created, by constantly proving itself as a desirable personal possession, by continually passing the test that its value in the hearts and in the personal possessions of the people is equal to the value placed upon it by governments.

The value of gold bullion at wholesale is largely artificial. It is created by the demands of a few governments for gold reserves, the majority of governments not being able to exercise any effective demand. But such demands as are actually effective are artificially created by rule-of-thumb proportions having no scientific validity. In the past five years the currency reserves of the United States have changed from a proportion of ten to one to a proportion of five to one. In the same period British demand for a gold reserve has increased from £150 millions to £750 millions and provision has been made to raise it to £1,000 millions, if necessary. At the present moment it is extremely difficult to say whether either the British or the American demands for gold are genuine, or are merely artificial demands for maintaining the price of gold at an unduly high level.

The only method of maintaining the natural value of gold is the free operation of the Law of Supply and Demand. The free operation of that Law means the maintenance of an intimate relationship between the wholesale and the retail demands and the full satisfaction of the retail demand. At present the retail demand (apart from the industrial usage of gold) has been artificially suppressed and does not function at all in the gold market. The Law of Supply and Demand, as far as gold is concerned, is now a truncated Law, in which the original and natural retail demand for gold, the individual demand which first gave gold its value as a precious metal, has been eliminated.

The natural price of gold is that which is the resultant of the free interplay of all the forces of Supply and Demand. Whether the price of £7 an ounce, 35 dollars an ounce, is a natural price or is too high or too low cannot be determined satisfactorily by a Demand which is limited to the artificially calculated requirements of the world's monetary authorities, most of whom have been for many years in too impoverished a condition to exercise any demand whatever.

The British and American monetary authorities can be released from the necessity of having to absorb in their reserves enormous quantities of gold which they do not want by this method. Relief can be obtained immediately by diverting the supplies of gold, so far as they are redundant, to satisfy the demand for gold where it is wanted most insistently. It would be less expensive for the British Treasury to satisfy the French demand for the storage of liquid funds by issuing golden sovereigns, than by accumulating extra gold reserves in order to issue extra pound notes to satisfy the same demand. It would also cause less monetary dislocation.

The demand for British sovereigns, if issued at the

prevailing market price for gold, would not be limited to France. It would be world-wide. The demand for gold coinage for storage has had little opportunity for satisfaction since the outbreak of war twenty-three years ago. There must be an incalculable latent demand from almost every country in the world, a demand equal to the total value of the "hot money" that is restlessly and disastrously wandering from country to country vainly seeking a medium in which it can settle down in complete contentment and rest. As Mr. Cohrssen states, it would soon be snatched up.

A doubt may arise whether this proposed solution of the gold problem would be too successful. Would the retail demand for gold in the form of sovereigns and half-sovereigns for storage, added to the demand for gold bullion for national gold reserves and for the requirements of the international currency, be greater than the gold supply? Would it be sufficiently great to cause a rise in the price of gold? It might. The experiment would have to be most cautiously and carefully supervised. Precautions would have to be taken that the demand for coinage should not compete unduly with the demand for gold bullion required for the restoration of national gold reserves. After all the care which has been given, and is still being given, to the adjustment of the world wholesale price level to the high price of gold, the equilibrium which is being effected with such difficulty should not be lightly disturbed.

Some experiment of this nature is necessary. The

problem of the future of gold is compounded of many subsidiary problems. Is the value of gold due to magic, or to superstition, or to the persistence of an obsolete tradition; or to a genuine demand for its services? Is the production of gold too great for the world's needs? Is the so-called over-supply of gold due to the numerous restrictions contracting the demand for its services; and especially to the virtual prohibition of the use of gold for its most ancient purpose, the storage of individual savings? Is the price round about £7 per ounce in harmony with the world's demand for gold, if that demand were allowed to operate freely? Are gold reserves a substantial, or an illusory, backing for national and international currencies? Is the immense superstructure of representative money based entirely on confidence in governments and banking systems; or has it a supplementary material basis on gold, a substance the value of which has been created by the democratic preferences of mankind as exercised for fifty centuries?

Until these subsidiary issues are decided, the problems raised by the large inflowing supplies of gold will baffle solution. It is essential that the value of gold shall not only be tested, but also maintained by the popular demand for its services. The recent panic, due to a belief that the price of gold might fall unduly, was mild compared with the panic which would arise if the present restrictions preventing the possession of gold by the people at large should create a popular doubt as to its value.

CHAPTER IX

A SIMPLE SOLUTION

"Thus, gold mines are of the greatest value and importance to civilisation. Just as wars have been the only form of large-scale loan expenditure which statesmen have thought justifiable, so gold-mining is the only pretext for digging holes in the ground which has recommended itself to bankers as sound finance; and each of these activities has played its part in progress—failing something better."—Mr. J. M. Keynes—*The General Theory of Employment, Interest and Money*, Chapter 10.

"A restoration of confidence in the future of gold might conceivably stimulate some renewal of the hoarding demand, and in this connection the increased enquiry in evidence at the end of last week may not be without significance."—City Notes, *The Times*, July 26, 1937.

Is the savage irony of Mr. Keynes justified? Is gold-mining merely a plausible pretext for digging holes in the ground? Is the cost of producing gold a wasteful expenditure of £250 millions per annum. Is the belief in the utility of gold in the twentieth century a superstition on the same level as the worship of the golden calf by a nomadic tribe in the Dark Ages? Is gold a fetish, or a valuable commodity? Is the City Editor of *The Times* correct in anticipating a renewal of the hoarding demand for gold; that is, for the exercise of liquidity preference in gold rather than in other forms of money?

The pre-War economic fallacy that money did not

matter is in danger of being succeeded by the equally mischievous fallacy that gold does not matter. In the spring of 1937 there was an acute monetary panic, followed by severe dislocations in the Stock Exchanges and Money Markets of the world, due to the belief that the American price of gold was to be reduced. That panic, and the ensuing dislocations, financial and industrial, would be insignificant trifles as compared with the comprehensive overthrow of all the monetary systems of the world that would ensue if the dangerous doctrine that gold does not matter were to gain general acceptance. Should that fallacy be accepted as a truth, or transformed into a truth by the withdrawal of gold from the human contact which created its value, the superstructures of international currency and of representative money would inevitably collapse.

The opinion that gold does not matter is so far from being true that, for the twenty-three years since the outbreak of war until the present year, the demand for the services of gold has been greater than the supply. It was not till the year 1937 that supply threatened to exceed the artificially restricted demand for its services.

After the War the demand for the services of gold was so great in proportion to the supply of gold that it was withdrawn from circulation and concentrated for use as national gold reserves. This withdrawal was justified on the ground that gold was no longer required as a short-term internal medium of exchange. Representative money could perform the day-to-day functions of an internal medium of exchange as well as gold.

But gold was still required for other monetary functions, including the settlement of international balances, the financing of movements of capital from one country to another, and the guaranteeing of the liquidity of the immense stores of value embodied in corporate funds and held in various forms of representative money. These functions could be performed by gold reserves held in the form of bullion and stored in the vaults of the Central Banks.

But there was one monetary function that accumulated gold reserves in the form of bullion could not perform. They could not satisfy the liquidity preferences of the multitude of individuals who desired to store their savings in gold. As long as the quantity of gold available was insufficient to perform the two storage functions:

1. the guaranteeing of the value of corporate funds stored in forms of representative money,
2. the direct storage of funds held by individuals,

it was the obvious policy to concentrate all available supplies on the former service, the more important of the two.

The shortage of gold, enabling it to provide one only of these two important and necessary storage services, however, caused widespread injustice to individuals by limiting their freedom in the exercise of their liquidity preferences. It also caused great disturbance and dislocation to finance, industry and commerce, by creating a large amount of what is known as "hot money," that is, funds stored in monetary forms which do not possess the confidence of their owners.

This injustice and this dislocation were inevitable as long as the shortage of gold continued. But the shortage provided its own remedy. It caused such a rise in the price of gold (between 60 and 70 per cent) that it stimulated the production of gold to a volume more than sufficient for the limited monetary services it has provided for the past few years.

There is now enough gold, not only for national reserves to settle international balances, to finance international capital movements, and to guarantee the liquidity of the corporate funds stored in forms of representative money, but also to provide for the direct storage of funds to be held by individuals in gold coinage.

The diversion of gold to supply the demands of individuals for the personal storage of their funds is the only possible method to adopt in order to eliminate the dangerous menace of "hot money." The evil of "hot money" must be endured when gold is scarce in comparison with the total demand for its services. It can easily be eradicated when there are ample supplies of gold to enable the possessors of "hot money" to place their funds in the universally acceptable medium of gold.

The eradication of "hot money" necessitates the issue of coinage to provide gold in a convenient form for individual storage. The United States and Britain are at present the only countries with sufficient spare supplies of gold to provide coinage for this purpose. The United States is now using its surplus stores of gold to some extent to assist in replenishing the national gold reserves of China and Brazil and may extend its re-distribution of gold to other South American countries. But the liability of a possible heavy drain of gold will probably prevent the monetary authorities of the United States from relaxing the existing restrictions on the possession and export of gold.

The wide international outlook of Britain should induce her to consider how best to restore to the people of all countries the freedom to store their savings in the medium they have employed since the dawn of civilisation. The surplus supplies of gold arriving in Britain, in so far as they are not needed for international currency purposes or to restore the depleted gold reserves of other countries, can be profitably employed in being sold retail at the prevailing market price in the form of gold coinage. That is a far more serviceable use than merely adding them to the already adequate resources of the Exchange Equalisation Fund. It would be a return to the invaluable service performed by Britain in the nineteenth century in ensuring an equitable distribution of gold to all the nations of the world.

There are two courses open to Britain in dealing with the large and increasing supplies of gold at her disposal, in so far as they are surplus to her requirements:

1. to satisfy the small and restricted, but growing, demand for national gold reserves, assisted by the appropriate operation of forward exchange;
2. to supply, not only the wholesale demand for gold bullion for national gold reserves, but also the insistent retail demand throughout the world for gold coinage for individual storage purposes.

The first course is a long-term method of solving the gold problem. The second is much more speedy. It solves not only the problem of satisfactorily absorbing the abundance of gold now being produced, it solves a number of associated problems. It removes the injustice to individuals in not permitting them the freedom to exercise their liquidity preferences in the only medium which makes a strong appeal to them. It eliminates the evil of "hot money." It will maintain the price of gold, by the free operation of the Law of Supply and Demand, at a natural level which will prevent the severe deflation of all currencies. It will bring the present value of gold into close contact with the force which initially created its value, and maintained its value throughout the ages, the unfettered demand of the people. It will save the world from the cataclysm that will ensue if gold should lose its affection and its value in the hearts of the people, and leave the international currency and the superstructure of representative money without any material guarantee of their universal liquidity.

Gold has proved to be incapable of securing the stability of the purchasing power of money. But it is the only substance capable of guaranteeing the permanent and universal liquidity of representative money. A method of securing the stability of the purchasing power of money, independent of gold, has been discovered, and is now being applied. But that does not mean that the monetary services of gold in settling the international trade balances, in financing international capital movements, in securing the liquidity of the immense superstructure of representative money, and in providing a medium for the satisfaction of the liquidity preferences of the majority of mankind, can be abandoned.

The problem of "The Outlook for Gold," though serious and menacing, is not difficult. Its solution requires merely the abandonment of the scarcity measures adopted to meet the severe scarcity of gold which culminated in the year 1931. The withdrawal of gold coinage was the principal measure adopted to cope with an inadequate supply of gold. The continuation of that scarcity measure when scarcity has ceased and a period of superabundance of gold has arrived, is the source of many of the monetary evils from which the world is suffering.

The cure for the superabundance of gold is simple. It is the removal of the artificial restrictions which curtail the natural and insistent demand for the services of gold. The retail demand for gold in the form of coinage is a universal demand dating back to the beginning of history. It is as persistent now as ever. Its satisfaction will cost nothing. On the contrary, it would relieve both Britain and the

United States from a series of expensive and embarrassing problems.

The post-War monetary and financial dislocations have been largely due to the misunderstanding of the monetary functions of gold. It has been required to perform services beyond the range of its capacity. Its unique monetary quality is not the stability of its value, but its liquidity, its universal acceptability as a natural form of money in all the countries of the world. It possesses a natural and a universal liquidity, independent of the artificial liquidity conferred by governments on other forms of legal tender.

This natural acceptability and liquidity of gold is a monetary asset of a unique and incalculable value. It is the most purely democratic fact in the world. It has been created by the universal suffrage of mankind; not by the preferences of the people of one country, of one race, or of one civilisation, but by the acclamation of the peoples of all countries, of all races, and of all civilisations since the world began.

The utility of gold, the value of its monetary services, the confidence in the future that it inspires, are among the most valuable legacies inherited from the past. Some of these services, some of these utilities, can now be provided by more modern forms of money. But no modern monetary invention can provide for the countless millions of mankind that feeling of confidence in the future which has been attached to gold by the experience and the tradition of fifty centuries.

It is possible for the present generation to preserve

this priceless heritage from the past by utilising the present abundance of gold to permit its distribution to those who passionately desire its services and believe in its value. It is equally possible to lose this heritage by burying all gold, as it is produced, in the vaults of the Central Banks, where it will be divorced from the human contact which first created, and has since maintained, its value. The Outlook for Gold depends on the correct decision between these two alternatives.

INDEX